Silica-Based Materials for Advanced Chemical Applications

Dedication

This book is dedicated to my beloved father,
Francesco.

Silica-Based Materials for Advanced Chemical Applications

Mario Pagliaro

CNR, Istituto per lo Studio dei Materiali Nanostrutturati and Institute for Scientific Methodology, Palermo, Italy

RSCPublishing

ISBN: 978-1-84755-898-5

A catalogue record for this book is available from the British Library

Published by The Royal Society of Chemistry,
Thomas Graham House, Science Park, Milton Road,
Cambridge CB4 0WF, UK

Registered Charity Number 207890

For further information see our web site at www.rsc.org

Preface

"Sol–gel improves the quality of human life" says the masthead logo of the Israeli company (Sol-Gel Technologies Ltd) co-founded by the pioneer of sol–gel technology for obtaining molecularly doped silica glasses. Today indeed the millions of adolescents suffering from acne can finally treat the symptoms with sol–gel entrapped benzoyl peroxide without seeing their skin become further inflamed (Chapter 2). And if you think that *Acne vulgaris* is not a serious enough illness, now think of sol–gel entrapped cells of the *Taxus brevifolia* tree used to synthesize the powerful anticancer drug Taxol (Chapter 5); or consider the sol–gel entrapped cytochrome P450 enzymes at the basis of the MetaChip (Chapter 5) thanks to which compounds that merit further development as new drugs can be identified simultaneously and much faster than in the recent past.

Human life, furthermore, certainly benefits from a less polluted world, and here, again, sol–gel entrapped catalysts are, literally, able to have transferred to within their large inner porosity the whole chemistry of fine chemicals production. Think for instance of an innocuous easily handled orange powder called SiliaCat TEMPO (Chapter 5) that added to a mixture of alcohols at 0 °C with a modest excess of aqueous, cold bleach rapidly converts them into all those fragrances, vitamins, hormones and drugs made of carbonyl compounds.

And now think of the facade of a house painted with one of those new paints containing the sol–gel hybrid formulation recently commercialized by the world's largest chemical company (Chapter 4) that now will retain its fresh aspect for years.

Silica-Based Materials for Advanced Chemical Applications
By Mario Pagliaro
© Mario Pagliaro 2009
Published by the Royal Society of Chemistry, www.rsc.org

All of these—and the innumerable other—revolutionary applications that are finally reaching the market are obtained from a single class of materials, namely silicas and organically derivatized silicas.

****

The hydrolytic polycondensation of silicon alkoxides of general formula $Si(OR)_4$ or $R'_nSi(OR)_{4-n}$, where the non-reactive organofunctional group R' acts as a network modifier, is carried out in the presence of dopant molecules resulting in the formation of highly porous, reactive organosilicates whose applications span many traditional domains of chemistry.

If R' can react with itself or additional components (R' contains vinyl, methacryl or epoxy groups, for example), the result of the condensation process is a flexible network of inorganic oxide covalently bonded to organic polymers, namely a hybrid nanocomposite lacking interface imperfections. The properties of this hybrid nanocomposite are intermediate between those of polymers and glasses, and can meet unique requirements.

First demonstrated by David Avnir in 1984, the principle is as simple as it is potent. Due to the low temperature needed for the preparation of sol–gel matrices, almost all of the 18 million existing organic and bioorganic molecules that could not be doped in glass, because glass is prepared at elevated temperatures (about 1000 °C), can now be entrapped in sol–gel glasses.

Organic chemistry and biochemistry have merged with the chemistry of ceramics.

Perhaps not surprisingly, at the same time as the doping methodology was being developed, researchers in Germany introduced a process which allows a precise control of hydrolysis and condensation rate of organosilicon and other metal alkoxides (such as those of aluminium, titanium and zirconium), which is the basis for making ORMOCER materials commercialized as protective hard coatings for transparent plastics since 1988.

When in 1994 the US and European patent offices recognized that the doped sol–gel technology had been invented by David Avnir and colleagues, granting the Hebrew University of Jerusalem a series of patents covering the generic methodology for the preparation of sol–gel materials, and their use in various applications, the time was ripe for the foundation of the first sol–gel chemical companies.

And indeed since then a large body of new companies has been established in countries as far from each other as Finland, Australia, the

USA, Italy and Germany. These have developed in-house, unique sol–gel processes to produce an arsenal of silica-based materials for drugs release and screening, catalysis, sensing, chromatography, *i.e.* addressing differing application needs.

****

Writing a book in this burgeoning field runs two risks. The first is oversimplification, as there are entire volumes dedicated to single chapters of the present text (*e.g.* G. Kickelbick (ed.), *Hybrid Materials*, Wiley-VCH, 2006). The second is to produce another lengthy book aimed at disseminating research that, in the information overload era of the internet, would rapidly join all those ignored scientific texts (M. Reisz, Publish and be ignored, *The Times Higher Education*, 24 April 2008).

Instead, the aim is to provide a unified picture of the chemistry of functional silica gels. Hence, in place of a complete coverage of what has been done with these immensely versatile materials, an attempt is made to provide readers with an understanding of the principles behind the applications.

In other words, in the spirit of older scholarly publishing, the overall goal is to produce a readable, well-illustrated textbook in which elucidation of principles may ensure durability; and updated information on products, companies, markets and technology trends may render it a fresh reading.

Finally, publication of a book is inevitably linked to the times in which it is produced. And we are convinced that the present times are indeed right for such a book.

If the objectives have been achieved, then the book will not suffer from the obsolescence syndrome that makes a book outdated in the time lag between delivery of the manuscript to the publisher and the finished book's distribution.

****

Finally, readers will not find a single chapter dedicated to biogels— silica gels entrapping biologicals—but rather specific examples of their usages in different fields. As the development of biotechnology is based on the immobilization of biomolecules or micro-organisms onto solid substrates, these materials—those already developed and those that are being created—are making a reality of biotechnology that has been awaited for many years (D. Avnir, T. Coradin, O. Lev and J. Livage,

Recent bio-applications of sol–gel materials, *J. Mater. Chem.*, 2006, **16**, 1013). These materials deserve a thorough, consistent treatment that would exceed the scope of the present text.

This book should be useful to researchers and undergraduate students who carry out research in the field, and to managers and management consultants in the chemical industry who will gain a clear picture of what this technology is all about and how it can be used to solve their specific problems.

A section of my website (qualitas1998.net) features freely accessible additional teaching and communication materials (lecture slides, articles, links to companies and research groups, *etc.*). Readers are warmly invited to send their feedback: it will be used as a basis to improve future editions of the book.

Mario Pagliaro
Palermo

Contents

Silica-Based Materials for Advanced Chemical Applications
By Mario Pagliaro
© Mario Pagliaro 2009
Published by the Royal Society of Chemistry, www.rsc.org

Chapter 8   Strategic Aspects of Functional Silicas

Acknowledgements

Without Dr Rosaria Ciriminna's exceptional chemical talent, none of what we have achieved at Palermo's CNR would have been possible. I thank her for all the creative joint work, and for a unique style, that enabled the introduction of several new sol–gel functional silicas for a multiplicity of applications (and much else).

Starting in late 1996, the University of Jerusalem's Professor David Avnir, the pioneer of sol–gel doped materials, has been much more than a mentor. I thank him also for all he has done for me in all these years. The wonderful work of Sharon Marx with chiral sol–gels is impressive, and also, Sharon, the celebration of Hanukkah at your parents' home in Jerusalem in late 2001 will not be forgotten.

I am indebted to Professor Daniel Mandler, also from the University of Jerusalem, for his memorable talk on science in Israel, given at our institute in May 2007.

PhD student Giovanni Palmisano joined us in 2005 and is now completing a remarkable doctorate in chemical engineering at Palermo University. His talent, commitment and constancy of purpose will form the basis of a similarly great career in science.

Montpellier's ENSC director Joël Moreau and senior researcher Michel Wong Chi Man have taught us plenty of advanced materials chemistry. We thank them for a splendid collaboration and for their warm hospitality.

Working with the University of Padua's Professor Sandro Campestrini and Dr Massimo Carraro has been (and is) as fruitful as it is pleasant. And similarly joyful is the research with Professor Laura M. Ilharco and Dr Alexandra Fidalgo at Lisboa Polytechnic. We thank the University of Poitiers' Professor Eric Le Bourhis and Madrid's CSIC senior researcher Marisa Ferrer for rewarding shared work on sol–gel thin films and on photochemical and biomaterials, respectively.

Cooperation on the use of sol–gel marine antifouling with Professor Sebastiano Tusa and Dr Pietro Selvaggio, of Sicily's Superintendence of the Sea, is gratefully acknowledged along with that of Professors Michael Detty and Frank V. Bright of the University of New York at Buffalo.

The R&D director of SiliCyle, Dr Francois Béland, has been pivotal in guiding the development of the SiliaCat series of sol–gel entrapped catalysts. And I would like to recall here also his and Hugo St-Laurent's visit to Palermo in October 2007.

I wish to express my appreciation to the academic colleagues who have exchanged information relevant to the book. They are the leading researchers in many of the fields discussed therein and this work owes much of its relevance to their scientific effort. I thereby acknowledge the assistance of Dr Christophe Barbé (Ceramisphere), John Berg (University of Washington), Boris Mahltiz (Gmbu, Jena), and Paul Ducheyne (University of Pennsylvania).

Finally, I am grateful to Dr Arjan E. J. de Nooy (now at Rijswijk office of EPO) for all he has taught me about chemistry and scientific writing; but also, and above all, about love for life and friendship.

RSC Publishing's Katrina Harding and Merlin Fox were instrumental in producing this book.

About the Author

Mario Pagliaro (born in Palermo in 1969) is a research chemist and management thinker based in Palermo at Italy's CNR, where he leads a research group and the new Institute for Scientific Methodology. His research focuses on the development of functional materials for a variety of uses and operates at the boundaries of chemistry, biology and materials science. Between 1998 and 2003 he led the management educational centre, Quality College del CNR, using the resulting income to equip his laboratories and establish a research group which currently collaborates with researchers in ten countries.

Mario holds a PhD in chemistry from Palermo University (1998), the topic of his thesis being the selective oxidation of carbohydrates; mentors were David Avnir in Jerusalem and Arjan de Nooy in the Netherlands. He has also studied and worked in France and Germany. In 2005 he was appointed *Maître de conférences associé* at the Montpellier Ecole Nationale Supérieure de Chimie. Between 1993 and 1994 he worked in the Netherlands, initially at the Rijks Universiteit, Leiden, and then at the TNO Food Research Institute in Zeist. In 1998 he was with Michel Vignon at the Grenoble CNRS, and in 2001 he joined Carsten Bolm's research group at Aachen Polytechnic.

Mario has co-invented a number of novel technologies, some of which have been commercialized. He is author of the management books *Scenario: Qualità* and *Lean Banking* and co-author of *The Future of Glycerol* (RSC Publishing, 2008) and *Flexible Solar Cells* (Wiley-VCH,

2008). He is the author of three international patents and a large number of scientific papers. Since 2004 he has organized the prestigious seminar "Marcello Carapezza" and chairs the organizing committee of the 10th edition of the FIGIPAS meeting in inorganic chemistry to be held in Palermo in 2009.

CHAPTER 1

Functionalized Silicas: the Principles

1.1 Functionalized Silicas

The formation of a sol–gel functional silica gel takes place by hydrolytic polycondensation of suitable precursors in the presence of a dopant in solution. An example is given in the following (unbalanced) equation:

$$CH_3Si(OCH_3)_3 + Si(OCH_3)_4 + H_2O \xrightarrow{\text{dopant}} \text{dopant}@[(CH_3)SiO_nH_m]_p \quad (1.1)$$

All of the oxygen in the final solid comes from the water molecules which cause hydrolysis of the alkoxide. In general, the hydrolysis process enables control of the monomer → oligomer → sol → gel → xerogel (dry gel) transition at a molecular level.[1] The values of m, n and p are in fact dictated by several factors, including the concentration of H^+ or OH^- employed as catalysts, co-solvent, the existence of additives like surfactants, the water:silane ratio $(=r)$, temperature, drying method and even the size and shape of the final product that can be obtained as thin film, powder, capillary, monolith, *etc.* Up to the xerogel stage, the low temperatures employed in the process enable one to close the traditional gap between organic and ceramics chemistry, and the chemistry and physics of organic molecules becomes applicable within ceramic matrices.[2]

In general, these oxides show excellent optical quality including high transparency in the visible region that allows fluorescence as well as charge separation processes. Applications are numerous and range from

Silica-Based Materials for Advanced Chemical Applications
By Mario Pagliaro
© Mario Pagliaro 2009
Published by the Royal Society of Chemistry, www.rsc.org

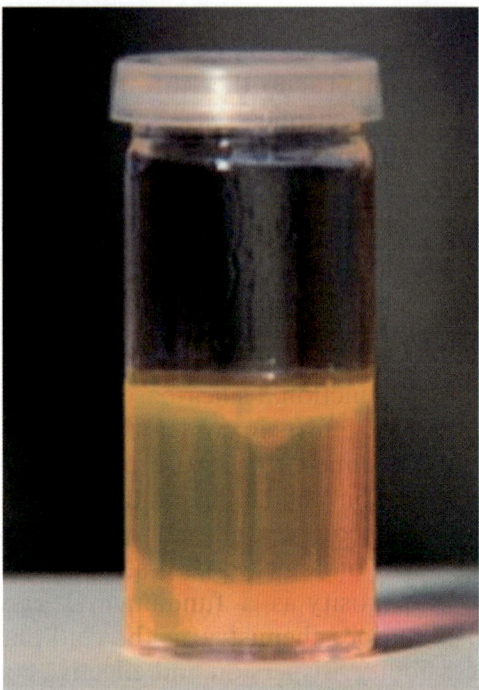

Figure 1.1   A silica sol precursor doped with a coloured organic molecule.

highly sensitive photochemical sensors to photochromic glasses. The enormous versatility of the sol–gel process exemplified in the choice of process parameters affords a potentially enormous class of doped materials, either as glassy solids (amorphous or periodic) or as crystalline solids in which the host–guest interaction can be tailored.

Doped sol–gel silica oxides (Figure 1.1), and organically modified silicates (ORMOSIL)[3] in particular, are thus used in a number of impressive applications that range from tailored organic light-emitting diodes (OLEDs) of enhanced durability and efficiency to promising efficient delivery of genes for gene therapy and fast drug assessment for toxicity; from self-ordered silica helices to highly sensitive photochemical oxygen sensors; and to "biochemical reactors" made up of entrapped enzymes, whole cells and even bacteria. In brief, doped ORMOSIL are functional materials with a multifaceted and exceptional chemistry, which are repeating the revolution that plastics caused in the 1940s and eventually finding a number of commercial, practical applications.[4]

Amongst other advantages, these materials offer the opportunity to utilize in a positive way geometric imperfections. In other words, similar to what happens in nature, complicated structures can be constructed with the sol–gel process allowing "correlations and disorder to compete and to come to terms with each other through an optimal solution".[5] What are thus the physico-chemical bases originating the ORMOSIL's chemistry superior performance in so many applications? How, furthermore, shape and structural effects in silica-based funtional materials are capable to dictate function?

1.2   Shapes Dictating Function

In a variety of powerful functional silica-based materials shape controls function and utility. Large efforts in contemporary research in materials science and biology are aimed to prepare materials with functionally powerful shapes, based on the understanding of the constructional processes that give rise to complex inorganic structures under the mild, wet conditions typical of biological processes. One general finding of these studies is that porosity is a fundamental part of any nanostructured materials that does chemistry, as the void phase – *i.e.*, nothing – ensures both accessibility, dispersion and effective confinement of any entrapped molecules. To paraphrase Davis, beyond the atoms and molecules that define the porous space, the challenge in the field of porous materials aims to control their *shape*. In other words, void space and deliberate disorder are used as design components, as disorder and geometrical imperfection of solid structures have long been known for their unique relevance to heterogeneous chemical processes. Eventually, the overall objective is to develop what has been called "a chemistry of form" in the laboratory.

Silica-based materials obtained by the sol-gel process are perhaps the most promising class of functional materials capable to meet such a grand objective. In the sol-gel process liquid precursors such as silicon alkoxides are mixed and transformed into silica via hydrolytic polycondensation at room temperature. Called "soft chemitry" or *chimie douce*, this approach to the synthesis of glasses at room temperature and pressure and in biocompatible conditions (water, neutral pH) has been pioneered by Livage and Rouxel in the 1970s, and further developed by Sanchez, Avnir, Brinker and Ozin.

These and several other researchers extended the methodology with the aim to widen functionality, using dopant molecules and silicon precursors derivatized with organic moities giving place to a vast class of hybrid organic-inorganic organosilica nanocomposites capable to meet numerous, advanced requirements in fields as diverse as catalysis,

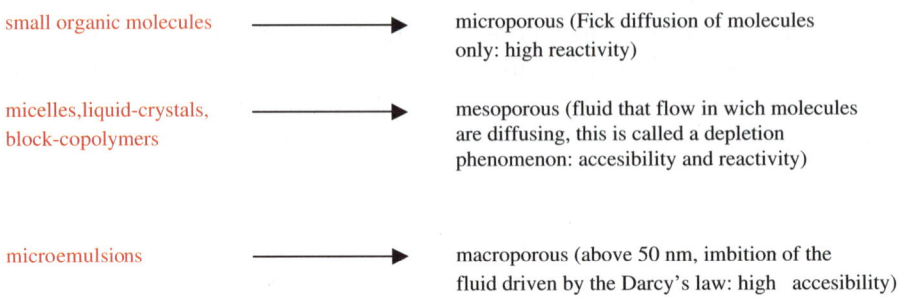

Scheme 1.1   How organic templates control porosity of materials (and consequences from a functional, reactivity viewpoint).

chromatography, surface coating, sensing, drug release and bio-technology. In general, from periodic mesoporous organosilicas (PMO) to Brinker's evaporation-induced self-assembly (EISA), sol-gel processing is coupled to molecular self-assembly as a simple, general means to prepare porous and composite nanostructures. Hence, by generalizing into chemistry the biominerals growth principles elucidated in the late 1910s, one may recognise how organics, and especially soft matter (lyotopic mesophases, foams, emulsions and beyond), are used to template all types of porous materials (Scheme 1.1).

Such emerging approach to functional materials has been named by Ozin "nanochemistry", and refers to a basic chemical strategy for making nanomaterials using molecular or nanometre scale building blocks (with a wide range of shapes, compositions and surface functionalities) that are further chemically processed to organize into structures serving as tailored functional materials. Indeed, solid state synthesis strategies in materials preparation are rapidly being supplanted by molecular methodologies, particularly the self-assembly of materials with structures that mimic the complexity of those observed in Nature. Almost inevitably, then, concepts such as anisotropy or symmetry become key parameters when considering "form" effects on the chemistry of these functional materials.

1.3   The Nature of Sol–Gel Entrapment

In 1995, commenting on the physical nature of the sol–gel entrapment of molecules in porous oxides, the inventor of sol–gel doped materials (first reported in the *Journal of Physical Chemistry* ten years before)[6] emphasized how it was "really remarkable to see how many applications of the entrapment have been reported, without fully understanding the picture at molecular level".[7]

Today of course this understanding has evolved and we have a broader picture of the factors governing the chemical behaviour of these materials, whose mild preparation conditions were soon shown to be compatible with the effective entrapment of biomolecules (with no loss, and often with enhancement, of biological activity) opening the route to the merger of chemistry, biology and materials science (Figure 1.2).[8]

Encapsulation of molecules into the inner porosity of a sol–gel matrix affords unprecedented molecular dispersion in a solid phase.

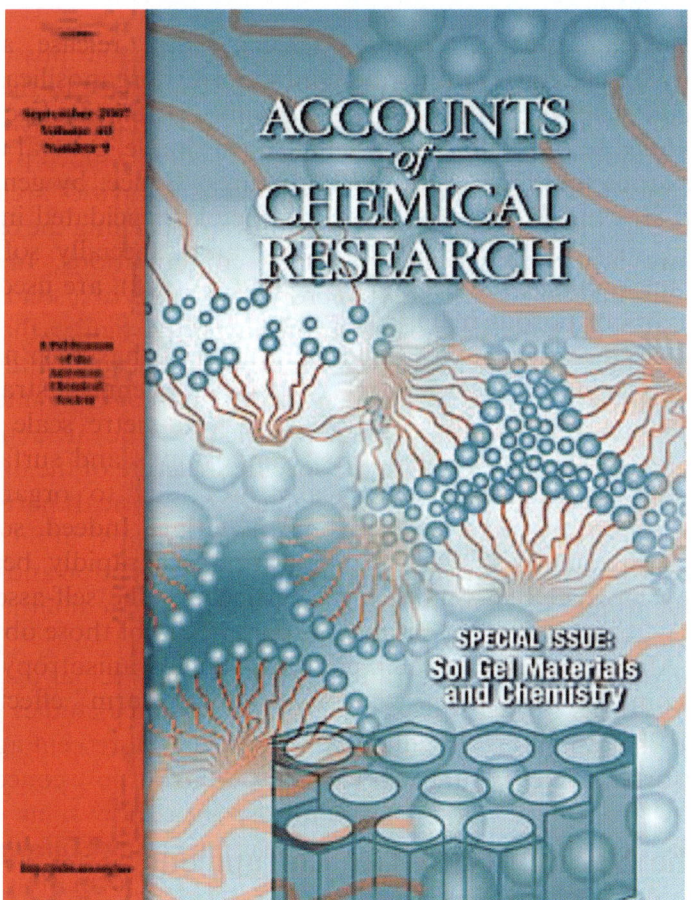

Figure 1.2   September 2007 cover of a leading international chemistry journal dedicated to sol–gel materials shows mesostructured silicas derived from the cooperative assembly of soluble sol–gel precursors and organic surfactant molecules. (Reproduced from acs.org, with permission.)

The question of dopant aggregation in sol–gel cages emerged in 1984 in the very first report of molecular sol–gel entrapment describing the photophysical behaviour of Rhodamine 6G confined in a SiO_2 matrix: whereas it was originally assumed that single-molecule caging was taking place, it was later established that at the high concentrations (of the order of several $mmol\,g^{-1}$) typical of doped sol–gel applications some dopant aggregation was actually occurring.

The physical and chemical properties of the entrapped dopants are generally retained. Yet, the efficient isolation of one molecule from another and the active role played by the sol–gel cage, for instance in dictating accessibility, gives place to a vast *new* chemistry and physics of sol–gel entrapped molecules which largely encompasses and goes beyond traditional solution chemistry.

Thus, for example, new optical applications become possible since entrapped excited molecules cannot diffuse (giving place to the thermal energy dissipation typical of molecules in the liquid phase) such as in the case of photoinduced electron transfer in SiO_2-co-entrapped pyrene and methylviologen.[9] Or, in catalysis, much higher selectivities are achieved in delicate organic syntheses as the reactants approaching the entrapped catalyst are forced to assume preferred configurations. In general, confinement of a molecule in solid microporous cages of a solid modifies the electron energy of the molecule, altering in particular the frontier molecular orbital energy. Moreover, entrapped molecules interact with the matrix and its surface by covalent or by non-covalent interactions (van der Waals interaction, π-stacking, electrostatic attractions and hydrogen bonding) depending on the specific structure of the matrix and of the dopant molecule and on the (chemical or physical) nature of the entrapment. For example, dyes physically encapsulated in SiO_2 glasses normally show a red shift in the positions of the absorption and emission spectra due to interaction of the dye molecules with the internal surface of the porous matrix.

Typically, at dopant concentrations $> 10^{-3}$ M molecular aggregation starts to occur as shown for example by fluorescence studies of silica-entrapped rhodamines (rhodamines tend to form fluorescent aggregates at the adsorbed state; Figure 1.3).[10] Hence, whilst at low concentration neither the absorption nor the excitation spectra show signs of aggregation, the excitation spectra of xerogels of increasing Sulforhodamine B (SB) concentration clearly show (Figure 1.4) the formation of fluorescent J-dimers.

High dopant loads (for instance, starting from a sol molar ratio composition TEOS:PhTES:dopant = 1:1:0.4 (TEOS, tetraethylorthosilicate; PhTES, phenyltriethoxysilicate)) are useful in photophysical

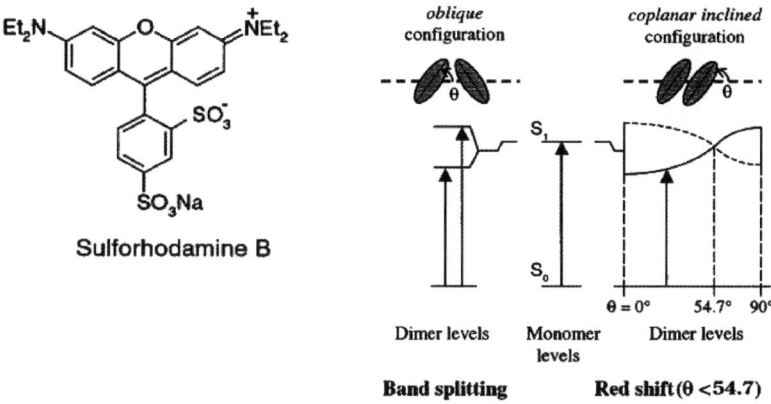

Figure 1.3 Molecular structure of Sulforhodamine B and exciton band energy diagram for molecular dimers with oblique and coplanar inclined transition dipoles. (Reproduced from ref. 10, with permission.)

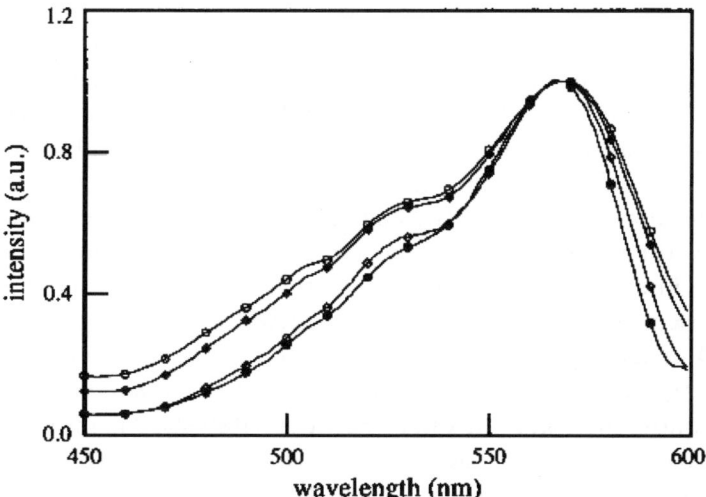

Figure 1.4 Excitation spectra ($\lambda_{em} = 650$ nm) of SB-doped xerogels of increasing ($\bullet \rightarrow \circ$) SB concentration. (Reproduced from ref. 10, with permission.)

applications where enhanced absorption is sought such as in the case of the UV protective coatings made of transparent phenyl-modified silica films doped with entrapped 2,2-dihydroxy-4-methoxybenzophenone protecting organic materials from light damage.[11] In the latter case, the use of an organically modified silica matrix enhances the solubility of the UV absorber in the matrix and allows the preparation of highly

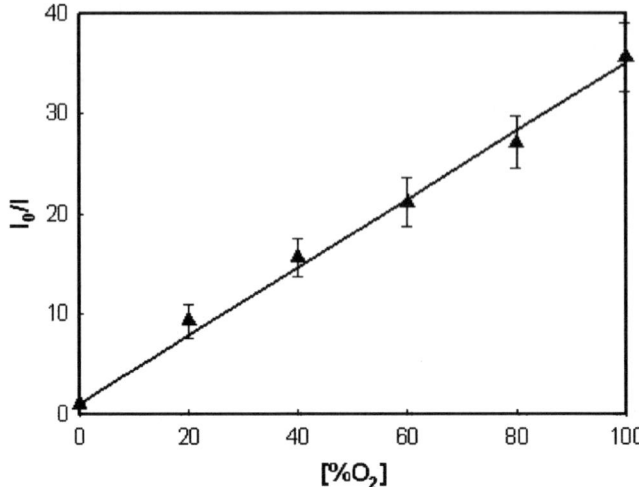

Figure 1.5   Stern–Volmer plot for a $[Ru(dpp)_3]^{2+}$-doped TFP-TriMOS:*n*-propyl Tri-
MOS-based sensor film. (Reproduced from ref. 44, with permission.)

loaded coatings. We recall here briefly that in SiO_2 samples prepared
from TEOS or tetramethylorthosilicate (TMOS), the surfaces of the
pores in the resulting matrix consist mainly of uncondensed OH groups,
which confer a very polar environment on the pore ($57.9\,kcal\,mol^{-1}$ in
the Reichardt $E_T(30)$ scale). Incorporation of R groups into the structure dramatically decreases the cage polarity with large organic groups
hindering further the influence of the residual OH groups at the pore
surfaces. The organic groups are located at the cage's interface, with
profound consequences for the homogeneity of the entrapment and the
chemical reactivity of the resulting material. The latter is indeed a
property that has large effect on the ability of doped sol–gel materials to
work as photochemical sensors or efficient catalysts. This is immediately
revealed by the first-order decay profiles of entrapped dyes[1] (Figure 1.5)
as nonlinear kinetic plots arise when the dopant chromophore report
simultaneously from more than one microenvironment that exhibit
different chemical properties.[12]

Furthermore, silica-entrapped molecules are physically and chemically *protected*. For example, organic fluorophores encapsulated *via* a
sol–gel in ORMOSIL nanoparticles (20–30 nm in diameter) become
20 times brighter and more photostable than their constituent fluorophore due to the cage protecting the fluorophore from bleaching due to
oxygen dissolved in the solvent.[13]

Figure 1.6 shows indeed evidence that the silica nanoparticles are
largely impermeable to solvent as the spectra show little spectral red

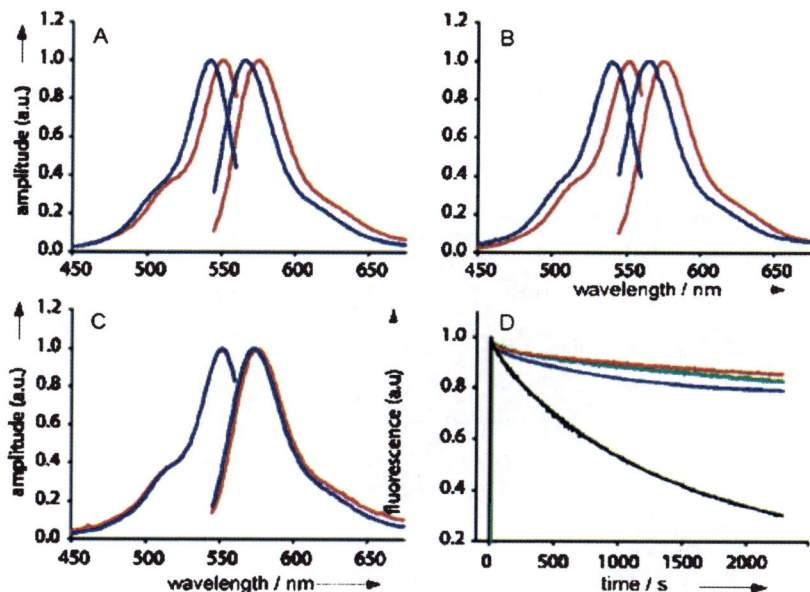

Figure 1.6 Solvent accessibility and photobleaching behaviour of nanoparticle synthesis intermediates. (A–C) Excitation and emission spectra of nanoparticle intermediates ((A) TRITC; (B) core; (C) core–shell) in ethanol (blue) and water (red). (D) Photobleaching behaviour of nanoparticle intermediates (blue, TRITC; green, core; red, core–shell) and fluorescein (black). All curves in (A)–(D) are normalized by the peak values. (Reproduced from ref. 13, with permission.)

shift in the excitation and emission spectra upon solvent exchange from ethanol to water.

Such good protection afforded by the sol–gel cage has tremendous consequences for biochemical applications using entrapped enzymes. For example, by entrapment in (surfactant-modified) silica sol–gel matrices, alkaline phosphatase remains functioning in extreme acidic environments, and acid phosphatase works smoothly in extreme alkaline environments.[14] This is due to the unique fact that large pH changes in very small local environments—such as the free space between the outer surface of the protein and the silica surface of the cage—actually mean very small variations in the actual number of protons (Figure 1.7).

Supposing that the water layer is a small reservoir of 100 water molecules, and that the external pH is 0, then the hydronium ions penetrate that reservoir until equilibrium is reached and a nominal "pH = 0" is also obtained in the layer's volume. From the point of view of the protein, this means that the protein gets protonated by only two protons ("pH = 0" means 2 moles H_3O^+ for each 100 moles water)

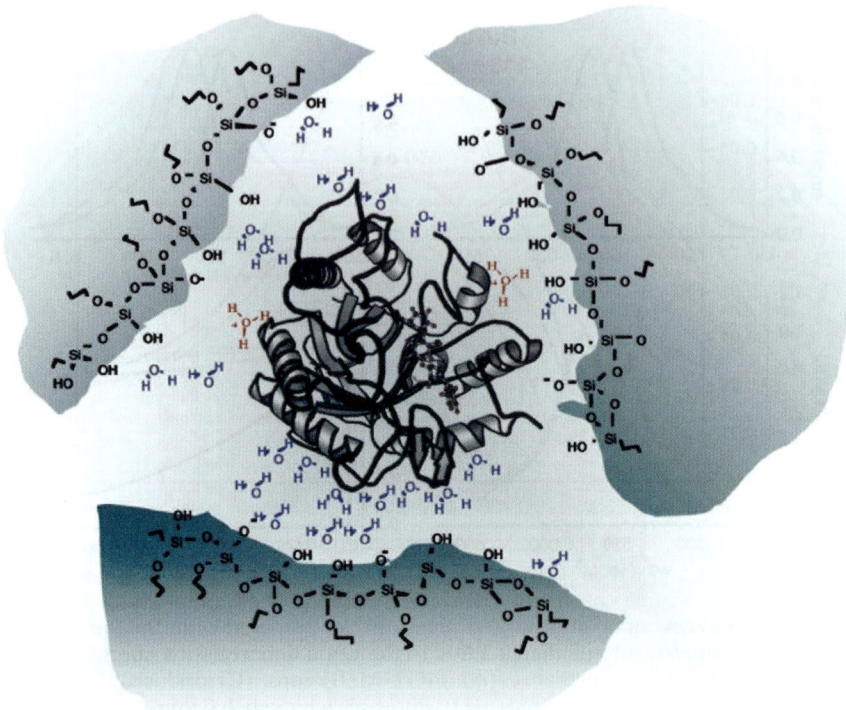

Figure 1.7   Schematic view of an entrapped enzyme with a few water molecules inside
                  a cage, two of which are protonated: the nominal "pH" is very low.
                  (Reproduced from ref. 14, with permission.)

which are enough to compensate for the extreme pH gradient while
clearly posing no stress at all for the encapsulated protein.

The effects of the hydrophilicity–lipophilicity balance (HLB) of the
sol–gel matrix on the chemical *reactivity* of the resulting material are
large and mostly due to the enhanced cage flexibility.[15] Referring to the
photophysical behaviour of ORMOSIL-entrapped naphthopyrans
(Figure 1.8), the shape and larger size of the isobutyl groups is also
responsible for the increased flexibility as compared with methyl and
phenyl groups, which facilitates the movement of the photochromic
molecules inside the pore resulting in faster isomerization kinetics.[16]

Getting back to photochemistry, photochemical reactions are kineti-
cally controlled conversions ubiquitous in nature where phenomena far
from equilibrium are the rule, rather than the exception. They are
generally categorized into two groups: those from equilibrated excited
molecules (with reactive species with lifetimes usually in nanoseconds or

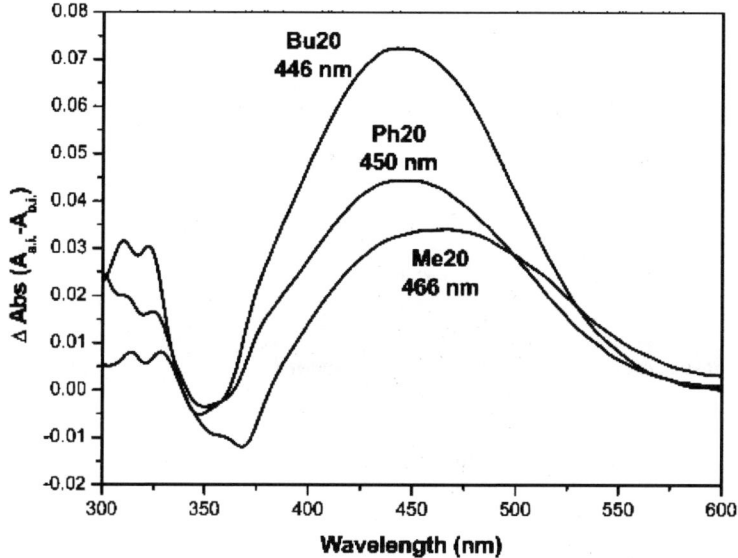

Figure 1.8 UV–visible spectra of ORMOSIL-embedded naphthopyrans with different modifying groups show that samples prepared with iBu and Ph groups
lead to more significant shifts to the UV compared with samples prepared
with the same relative amount of Me groups. (Reproduced from ref. 16,
with permission.)

microseconds); and those from short-lived unequilibrated molecules
(with short-lived vibrationally excited species) in which reactivity is
relatively insensitive to minor environmental perturbation.

In the early 1960s it became evident that the reaction environment had
an important role in dictating the course of photochemical conversions
acting on the course of the relaxation processes and stabilizing photo-
products.[17] A constrained medium such as that of a porous matrix or a
micelle provides the restricted environment to stop any bimolecular
processes that could lead to degradation of products. These effects,
however, are subtle. For instance, confinement of a molecule within a
host instead of leading to inhibition of reactions of the trapped substrate
often results in enhanced reactivity and selectivity because confinement
does not mean steric inhibition of *all* motions of the entrapped host
molecule which may eventually enjoy less restriction of some motions
than in common solvents.

Remarkably analogous findings have been established in recent years
from the study of a number of catalytic species entrapped in sol–gel
glasses. In particular, molecular entrapment in hybrid organic–inorganic
ceramic matrices such as organically modified silicates resulted in

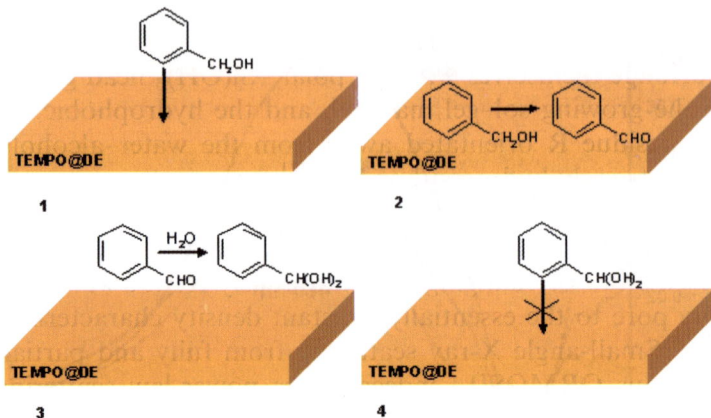

Figure 1.9 Alcohols are oxidized at the inner surface of TEMPO@DE (1 → 2). But not so hydrated aldehydes (3 → 4) which cannot enter the pores due to the hydrophobicity of the material's surface. (Reproduced from ref. 20, with permission.)

enhanced reactivity to transition metal, organo- and enzymatic catalysts, providing clear examples of heterogeneous catalysts in which the solid organic–inorganic surface participates actively in the reaction mechanism.[18]

For example, consider organic reactions in water. Beyond a general methodology for carrying out catalytic conversions in H_2O mediated by doped ORMOSIL in the presence of a modest amount of surfactant,[19] another recent method for the waste-free oxidation of alcohol affords high yields of commercially valued carbonyl compounds in water with complete selectivity and remarkable stability.[20]

The method uses a simple electrode made of a thin film of sol–gel organosilica doped with nitroxyl radicals deposited on the surface of an indium tin oxide (ITO) electrode. Thus, whereas in water benzyl alcohol is rapidly oxidized to benzoic acid, the use of the hydrophobic sol–gel molecular electrode TEMPO@DE affords benzaldehyde only (Figure 1.9), with an unprecedented purity, which is highly desirable for the fragrance and pharmaceutical industries where this aromatic aldehyde is employed in large amounts.

1.4 The Nature of the Sol–Gel Cage

Normally, the organic groups of ORMOSIL are located at the cage interface. The hydrolysis of organosilanes is slower compared to fully hydrolysable silicon alkoxides, and the slowly generated R–Si(OH)$_3$

monomers rapidly condense in micellar-like structures typical of the very early stages of the sol–gel process. These hydrolysed monomers tend to arrange themselves with the polar –Si(OH)₃ head groups at the front of the growing sol–gel material, and the hydrophobic non-polymerizable residue R orientated away from the water–alcohol solvent interfacial (strongly hydrogen bonding).[21]

As a result, the pore boundary (interface) of the resulting ORMOSIL has a "fuzzy" nature, *i.e.* the density varies continuously at the pore boundary instead of changing discontinuously from a value of zero in the empty pore to the essentially constant density characteristic of the bulk SiO_2. Small-angle X-ray scattering from fully and partially derivatized porous ORMOSIL, indeed, gives power-law scattering exponents of magnitude greater than 4, with the magnitudes of the exponents increasing with the alkyl chain length and with the degree of surface derivatization.[22]

Early analysis of absorption and emission spectra of a laser dye (R&G) in sol–gel SiO_2 indicated that the polarity of the microenvironment of the dopant is high and due to hydroxyls, although less polar than water.[1] The possibility to freely tailor the HLB by varying the amount and the nature of the organosilane employed as precursor is one of the most important features of sol–gel entrapped ORMOSIL. For example, by modifying the polarities of a series of organically modified sol–gel silicas doped with the solvatochromic dye ET(30) (2,6-diphenyl-4-(2,4,6-triphenyl-*N*-pyridino)phenolate) by copolymerizing, in various proportions, methyltrimethoxy-, vinyltrimethoxy-, propyltrimethoxy-, isobutyltrimethoxy- and phenyltrimethoxysilanes with TMOS, the various doped ORMOSIL exhibit a clearly distinct solvatochromic effect when in contact with organic solvents that makes these materials suitable as solvent sensors.[23]

Increasing organic content for ORMOSIL-entrapped lipase and perruthenate has a tremendous effect on the reactivity of the resulting catalytic material. A recent structural study[15] based on DRIFT spectroscopy aimed at investigating the origin of such large variations has shown that, in the absence of methyltrimethoxysilane (MTMS), the silica structure obtained by TMOS only is dominated by four-membered rings (the percentage of six-membered units in SiOMe is only ∼15%), but a *spectacular* increase in the fraction of six-membered rings (Figure 1.10) to 20, 56, 84 and 97% is observed as the MTMS content increases to 25, 50, 75 and 100%.

Conversely, the structure obtained when the precursor is only MTMS is almost entirely formed by the larger, less tensioned, six-membered rings, more able to accommodate the unreactive methyl groups. This

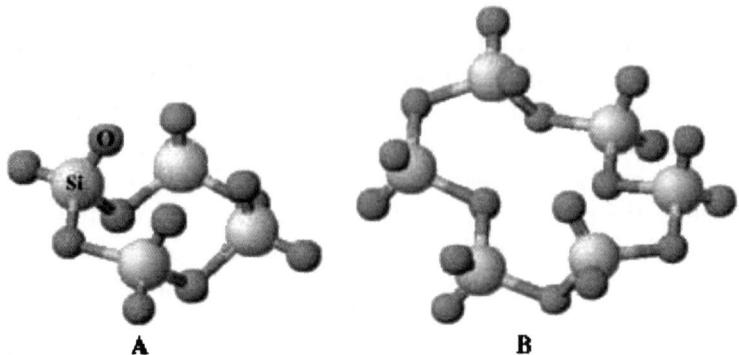

A   B

Figure 1.10   Schematic of the more common types of primary cyclic arrangements of
the structural units, SiO_4, in xerogels: (A) four-membered siloxane ring
$(SiO)_4$ and (B) six-membered siloxane ring $(SiO)_6$. (Reproduced from
ref. 15, with permission.)

explains to the "alkyl effect" observed in catalysis, where a catalytically
optimal 75% degree of alkylation is required for optimal catalysis: along
with enhanced hydrophobicity (promoting diffusion of the hydrophobic
reactant molecules within the porous network), enhanced cage *flexibility*
is needed to substantially enhance the reactivity of ORMOSIL-entrapped
catalysts in the liquid phase.

Quantification of the hydrophilicity of ORMOSIL may be given by
the contribution of dangling oxygen atoms relative to the silica structure
(% $Si-O_d$). Obviously, the *hydrophobicity* increases significantly for the
organically modified silica gels, and, although it is not very sensitive to
the modifier content, it reaches a maximum for 75% MTMS; but the
lipophilicity of the modified ORMOSIL is essentially due to the stable,
unreactive $Si-CH_3$ groups. DRIFT spectral analysis allows one to
conclude that the presence of the MTMS co-precursor affects the
structure *and* the HLB of the sol–gel catalysts. A rough estimate of the
HLB trend can be obtained from the ratio of the absorption peak areas:

$$HLB \approx [A(Si - O^-) + A(Si - OH)]/[A(O)C - H_3 + A(Si)C - H_3] \quad (1.2)$$

Again, this shows that, although modification with 25% of MTMS leads
to a significant decrease in catalyst hydrophilicity without major structural changes, an increase in the MTMS content does not affect
appreciably the catalyst hydrophilicity, but is responsible for a *gradual*
lipophilicity increase, and therefore for a decrease in the HLB, and for
striking structural changes in the silica network. Similar findings
obtained for isobutyltrimethoxysilane (BTMOS)-derived thin films that

exhibit a "discrete" change in polarity for films of greater than approximately 50% organic content are attributed to phase separation and/or the formation of micelle-like domains in the BTMOS-derived films.[24]

As mentioned above, the organic groups of ORMOSIL are normally located at the cage surface. However, if they are forced to migrate to the bulk of the material, even organically modified silicates suddenly become superhydrophilic.[25] This is what happens on heating a methyl-modified ORMOSIL foam above the temperature of such a transition (typically 400 °C) to thermally induce a contact angle change.

The equation

$$\cos \theta_r = f \cos \theta_s \pm (1 - f) \tag{1.3}$$

governs the behaviour of a drop sitting on a heterogeneous surface (θ_r is the contact angle of the liquid on a rough surface, θ_s the contact angle of the liquid on a smooth surface having the same chemistry as the rough surface, f is the fraction of the base of the drop in contact with the solid and $(1 - f)$ is the remaining fraction of the drop base). If the material has a high volume fraction of pores (such as in the case of mesoporous sol–gel organosilica foams), f will be small so the second component of the equation will dominate and one would expect the liquid to switch from very high (nearly 180°) to very low (nearly 0°) contact angles when the contact angle on a flat surface is varied by a small amount around the value of 90°; at this point sudden intrusion into the pores will occur (Figure 1.11).

1.5 Tuning Dopant Chemical Reactivity

While the sol–gel matrix can usefully preserve the properties of the dopant, the reverse is also true in that changing the sol–gel cage environment may

Figure 1.11 Phenolphthalein in water on methyl-modified silica foam heated to 390 °C (left) and 400 °C (right). (Reproduced from ref. 25, with permission.)

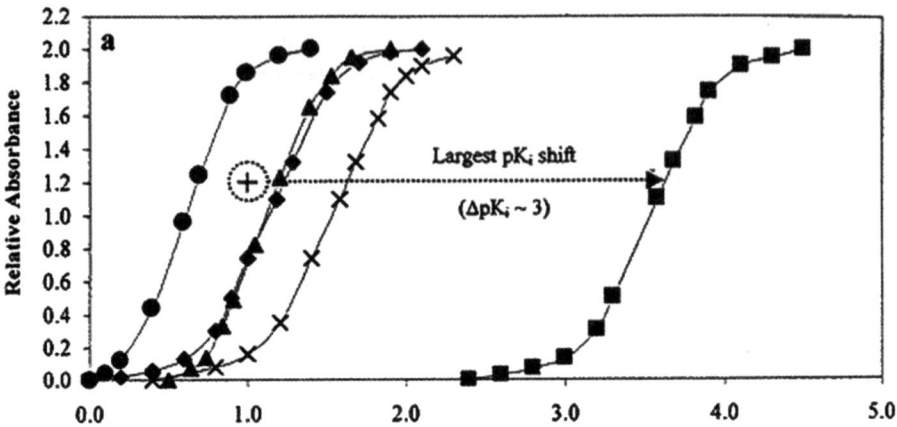

Figure 1.12 Photochemical titration curves of crystal violet co-entrapped in silica
sol–gel matrices with different surfactants, no surfactant (♦) and in
solution (×) show the impressive variations in the sensing properties for
the same entrapped dye. (Reproduced from ref. 26, with permission.)

well influence the properties of the dopant itself. This has been demonstrated by careful co-entrapment in a single sol–gel interphase of different
surfactants along with a dopant dye (crystal violet; Figure 1.12) as a
fluorescent pH sensor observing changes in the sensing properties of the
resulting materials from large magnitudes to delicate fine-tuning.[26] In
other words, different reactivities for the *same* molecule can be accessed by
tailoring the heterogeneous microenvironment of the molecule, thus
creating a library of reactivities from a *single* specific compound.

Instead of applying synthetic methods to alter chromophore reactivity,
this new way of controlling chemical reactivity involves choosing an
appropriate solid micellar system (from the available multitude) and
exploiting it to manipulate the chemistry of the entrapped compound. The
sol–gel matrix and the micellar solubilization, in fact, have a synergetic
effect. Their combination produces effects stronger and more tuneable
than in solution, so that a careful selection of sol–gel entrapped surfactants allows one to induce enormous changes in the dopant properties.

1.6 Tailored Structures

Controlling the sol–gel process at the molecular level implies the ability to
control   the   monomer → oligomer → sol → gel → xerogel   transition
affording the resulting porous oxide. Being soluble in common organic
solvents, undergoing rapid hydrolysis and being easily functionalized with
organic groups, silicon alkoxides such as $Si(OR')_4$ and $R_nSi(OR')_{4-n}$ are

conveniently used as sources of the monomers undergoing sol–gel hydrolytic polycondensation. The wide availability of organosilanes with phenyl, amino, carboxyl, *etc.*, functionalities affords an enormous class of ORMOSIL materials.

Consider a trialkoxysilane as co-precursor for an initial 1 : 1 mixture of alkoxysilane and organoalkoxysilane, and recall that in the case of ORMOSIL condensation takes place *only* through elimination of water[27] and not by alcohol formation. The mechanism involves hydrolysis and condensation reactions:

$$Si(OR')_4 + RSi(OR')_3 + 7H_2O \xrightarrow{\text{hydrolysis}} Si(OH)_4 + RSi(OH)_3 + 7R'OH \quad (1.4)$$

$$Si(OH)_4 + RSi(OH)_3 \xrightarrow{\text{condensation}} (OH)_3Si-O-SiR(OH)_3 + H_2O \quad (1.5)$$

The overall hydrolytic polycondensation reaction (unbalanced) can be written as

$$Si(OR')_4 + RSi(OR'_3) \rightarrow [R'SiO_nH_m(OR)_q]_p \quad (1.6)$$

The hydrolysis rate of organosilanes is a strong function of the size of the alkyl group and steric hindrance. Actually, however, these reactions never result in the formation of pure silica oxides, and it is precisely the fact that $m \neq 0$ and that a large number of unreacted silanol groups exist at the material's surface that gives rise to the impressive variety of chemical applications of doped silica xerogels.[2]

The resulting gel is chemically unstable, since its alkoxy (OR) groups are subject to further hydrolysis by the unreacted water. Consequently, the dry gel (xerogel) obtained by removal of the residual solvent is also unstable, since the remaining hydroxyl (OH) groups can further condense. The net result is a "living" material that undergoes structural modifications, even months after preparation, so that, for instance, an entrapped transition metal catalyst could migrate to the surface of newly formed porosity affording a solid catalyst several times more active compared to the freshly prepared material (Figure 1.13).[28]

In general, the structure of sol–gel materials evolves sequentially as the product of successive and/or simultaneous hydrolysis and condensation and their reverse reactions (esterification and depolymerization). Thus, in principle, by chemical control of the mechanisms and kinetics of these reactions, namely the catalytic conditions, it is possible to tailor the structure (and properties) of the gels over a wide range. For example, stable silica xerogels of tailored particle dimensions, pore morphology, density and porosity, from relatively

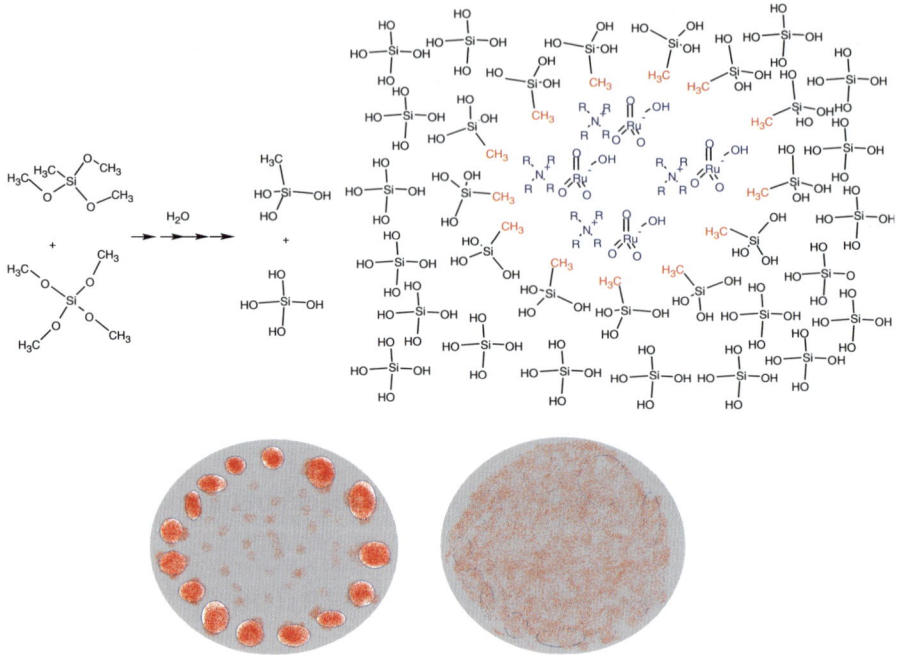

Figure 1.13   Pseudomicellar aggregates form in the early stage of the polycondensation of TMOS–MTMS catalytic mixed gels in which lipophilic TPAP (in red) tends to concentrate in the cores of inaccessible cages. Because with time the material continues to evolve, these segregated TPAP molecules migrate and disperse in newly formed inner microporosity resulting in 10-fold improvement in catalytic activity. (Image courtesy of Massimo Carraro.)

dense to highly porous, are easily obtained by carrying out the two-step sol–gel process—hydrolysis under acidic conditions followed by base-catalysed condensation—by tuning the acid and base contents in each step.

The silicon alkoxide is firstly hydrolysed in a strong acidic medium (in which the consumption of water due to silane hydrolysis is *very* fast, complete within minutes); the sol is then neutralized to promote condensation under a base-catalysed mechanism. In this two-step process, at the onset of condensation there are mostly hydroxyl groups bonded to silicon, favouring cross-linking. Furthermore, the base-catalysed polymerization mechanism also favours cross-linking, since the extent of the condensation reaction increases with the Si–OH acidity, which is stronger for the already more cross-linked clusters. The net result is that *larger* polymers grow at expense of smaller ones, leading to a highly cross-linked gel, which is essentially pure silica.

Finally, the versatility of sol–gel processing, and in particular the enormous flexibility associated with material fabrication, is well suited in particular for the preparation of thin films because of the short path length for reagent diffusion into the matrix.

1.7 From Disordered to Ordered Structures

In the absence of a templating agent, ORMOSIL are obtained in glassy, amorphous structures. However, highly ordered, organic–inorganic silica gels can be easily obtained in 'one pot' in the presence of a surfactant acting as a templating agent in which a micelle directs self-assembly of the polymeric siloxane precursors affording new materials whose *hierarchical* structure determines material properties and function.[29]

Such highly porous ordered silicas are made of amorphous silica walls spatially arranged into periodic arrays: the regular mesopores of monodisperse size (typically in the range from 20 to 100 Å in diameter) mimic the liquid crystalline phases exhibited by templates. The mesopore volume is usually greater than $0.7 \, mL \, g^{-1}$ with a narrow pore size distribution and extremely high surface area (up to $1400 \, m^2 \, g^{-1}$). As usual, these solids can be derivatized with a variety of organic groups to give mesoporous ORMOSIL either by post-synthesis grafting (reaction of organosilanes on mesoporous silica) or directly by sol–gel co-polycondensation of a tetraalkoxysilane and organoalkoxysilane in the presence of a surfactant.

Using a spray-drying process to disperse the precursor sol, for example, affords spheres of mesoporous ORMOSIL with functionalized surfaces made of highly ordered hexagonal domains (Figure 1.14).[30] However, co-polycondensation of organosilane with terminal groups such as $RSi(OEt)_3$ with TEOS is limited in scope since it tends to reduce the degree of long-range order and the mechanical stability for an organic content $> 20\%$ (in molar terms).[31] One can use as precursor the silica gel bridge-bonded silsesquioxanes of the form $(EtO)_3Si–R–Si(OEt)_3$ to afford a whole set of periodic mesoporous organosilicas: ORMOSIL with a high organic content (of $SiO_{1.5}R_{0.5}$ approximate stoichiometry) that are emerging as promising materials for applications such as nanocomposite hosts, sensors, catalyst supports and dielectric layers such as, for instance, low-k thin films made from mixtures of a silsesquioxane of the type $(C_2H_5O)_3Si–R–Si(OC_2H_5)_3$ with R = methene $(–CH_2–)$ and TMOS precursors.[32]

The dielectric constant (k) of the resulting ORMOSIL decreases with organic content with values as low as 1.9 for films entirely made of

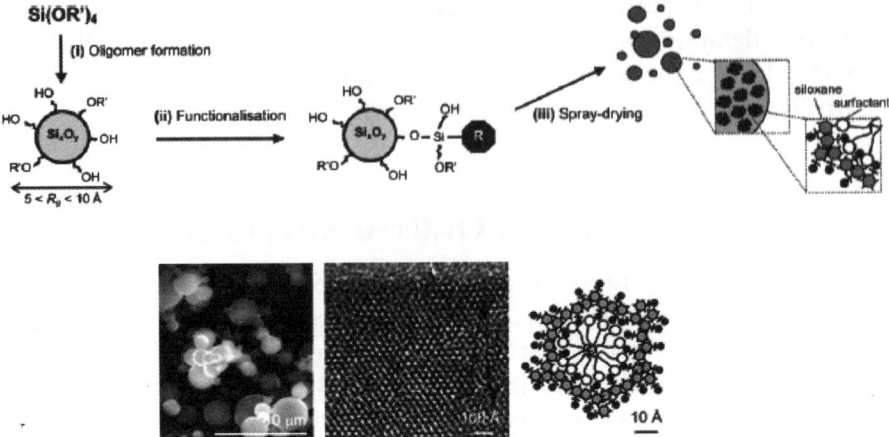

Figure 1.14 In a novel synthetic route made possible by the versatility of the sol–gel process, using spray-drying of the precursor sol affords spheres of mesoporous ORMOSIL whose surface consists of highly ordered hexagonal domains. (Reproduced from ref. 30, with permission.)

organosilica (*i.e.* using no TMOS). The films show good thermal stability and, because of their hydrophobicity (further enhanced upon calcination at 500 °C to cause a "self-hydrophobizing" bridging-to-terminal transformation of the methene to methyl groups with concomitant loss of silanols), excellent resistance to moisture.[21]

The merger of materials science and organic chemistry here is actually realized in its full potential and the number and type of materials obtainable are only limited by scientists' ingenuity. Hence, in the last few years a plethora of hybrid organic–inorganic materials of controlled molecular organization, nanoscale periodicity and macroscopic morphology have been introduced, recreating the hierarchical organizations of many natural materials and thus obtaining functional materials for new, advanced applications.[33] Structuring of bridged silsesquioxanes *via* cooperative weak interactions, for instance, leads either to impressive organosilica helices[7] (Figure 1.15) or to lamellar structures. In the former case, chirality is simply transcripted from the precursor molecule to the hybrid solid as the chiral molecules undergo self-assembly prior to sol–gel polycondensation.[7]

On choosing between ordered mesoporous organosilicas and related amorphous organosilica gels, a delicate *balance* between enhanced accessibility and restricted diffusion must be found. In heterogeneous catalysis applications, for example, confinement of the catalyst in the regular mesostructure generates a shape-selective catalyst with a large concentration of well-defined active sites.[34] However, this excludes from

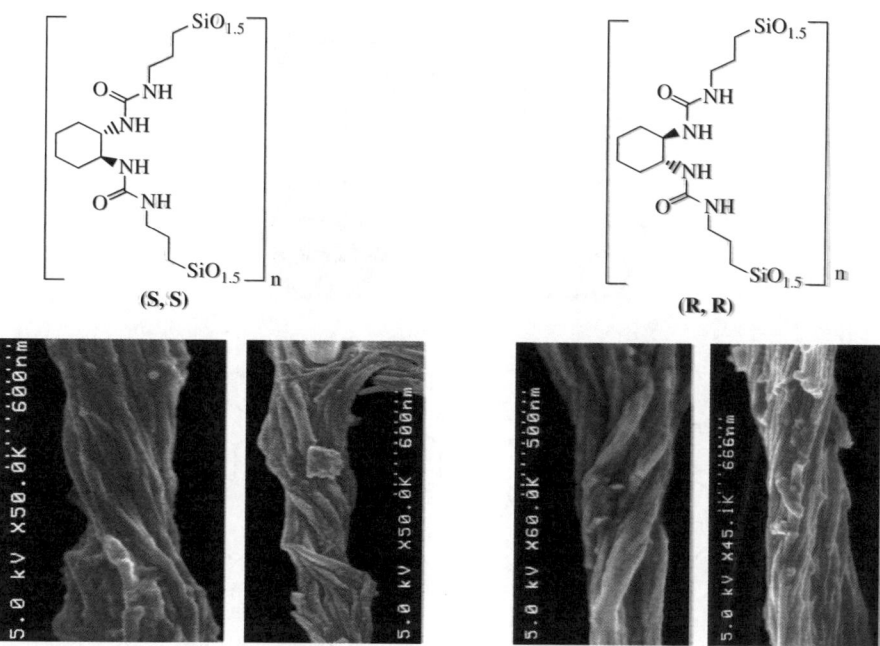

Figure 1.15 Helical hybrid silicas: chirality transcription from the precursor molecule to the hybrid solid. (Image courtesy of Prof. Joël Morean.)

conversion molecules whose size does not fit that of the pore. In contrast, amorphous glassy materials show a *distribution* of porosity that makes them versatile, *i.e.* they can be employed in the conversion of structurally different substrates.

For example, whereas the solid oxidation catalyst MCM-41-entrapped perruthenate can be used for the conversion of benzyl alcohols only, a similarly perruthenated-doped amorphous ORMOSIL is equally well suited for a variety of different alcohol substrates.[35] On the other hand, a uniform pore structure ensures access to the active centres, while in an amorphous material made of non-regular porosity hindered or even blocked sites can well exist (Figure 1.16), rendering the choice of the polycondensation conditions of paramount importance.

In general, the solid surface in molecularly doped organic-inorganic silica gels participates actively in the reaction mechanism, by:

- dictating access to the entrapped active molecule;
- providing a confined nanoenvironment that strongly differs from that experienced in solution;
- and reciprocally isolating, and stabilizing, the entrapped molecules.

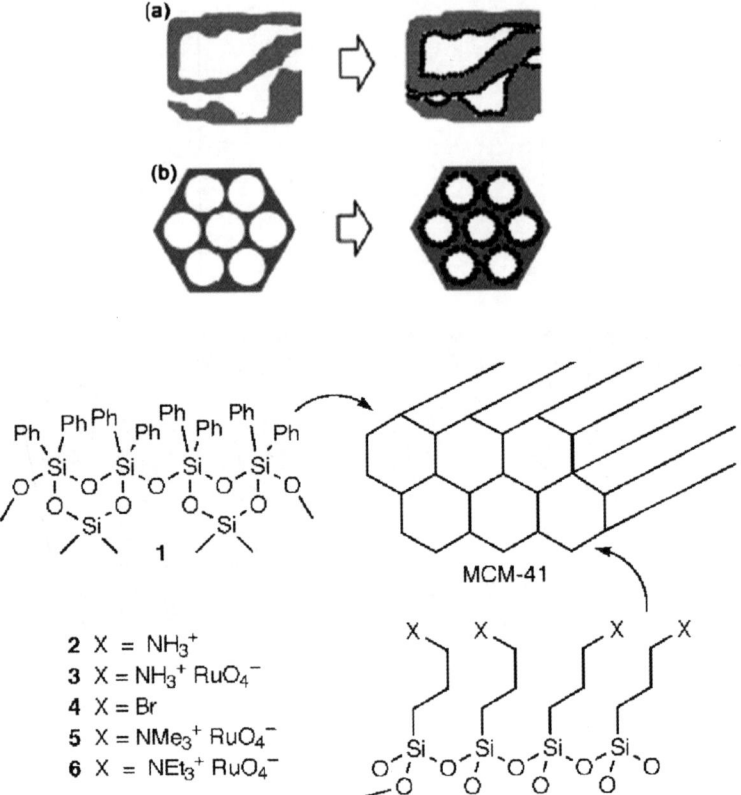

Figure 1.16 Schematic and comparative illustrative of grafting (a) an amorphous silica gel and (b) an ordered MCM-41 silica materials. Entrapment of oxidation catalyst perruthenate inside the channels of MCM-41 silica (*below*) results in a shape-selective aerobic catalyst, which may not be desirable from the viewpoint of catalyst versatality (Reproduced from ref. 39, with permission.)

Control over the material's shape at the nanoscale enables further control over reactants access to the dopant, and ultimately affords a potent means of controlling function which is analogous to that parsimoniously employed by Nature to synthesize materials with myriad function with a surprisingly low number of material's building blocks. A nice illustration is offered by the extrusion catalytic polymerization of ethylene within the hexagonal channels of MCM-41 mesoporous silica doped with catalyst titanocene.[36] The structure is made of amorphous silica walls spatially arranged into periodic arrays with high surface area (up to 1400 m^2g^{-1}) and mesopore volume >0.7 mL g^{-1}. In this case, restricted conformation dictates polymerization: the pore diameter

(27 Å) is much smaller than the lamellar length of ordinary PE crystals (~ 100 Å), thus the PE chains are prevented from folding. In this manner, extended-chain crystalline nanofibers of linear polyethylene with an ultrahigh molecular weight (6,200,000) and a diameter of 30 to 50 nm are formed.

Similarly, Hg(II) binding to thiol-functionalized mesoporous silica for which effective access to all the binding sites (100% of SH groups complexed with Hg(II) was achieved in micelle-templated mesostructures with pore diameters larger than 2.0 nm, whereas incomplete filling was always observed with corresponding amorphous silica-based adsorbents.[37]

A study of similar ORMOSIL materials prepared by self-assembly and co-condensation of mercaptopropyltrimethoxysilane (MPTMS) and TEOS in the presence of a cationic surfactant by varying the MPTMS content from 5 to 100% shows that the resulting porous solids exhibit clearly distinct structural order/disorder over different length scales (Figure 1.17).[38]

Total accessibility (expressed on the basis of a 1:1 S:Hg stoichiometry) was demonstrated for well-ordered materials containing up to 30% MPTMS. Less open structures characterized by a high degree of functionalization were subject to less-than-complete sorption capacities, while, however, reaching maximum loading values as high as 750 mg g^{-1}. The diffusion coefficients are strongly affected by both the structure and density of functional groups in the mercaptopropylsilane (MPS) sorbents.

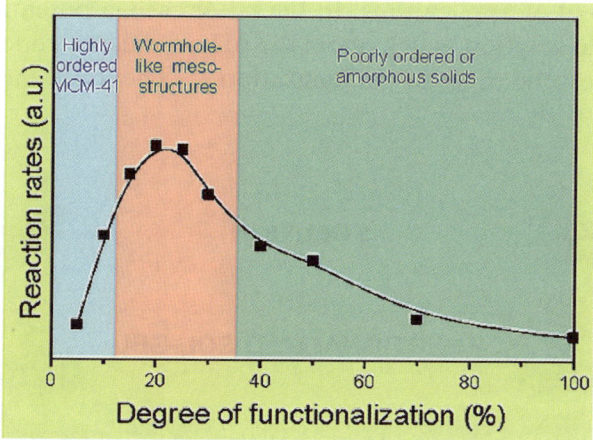

Figure 1.17 The sorption rates of mercaptopropyl-functionalized ORMOSIL depend on the relative long-range *versus* short-range structural order/disorder and their intrinsic hydrophobicity. Ordered materials with 20–30% alkylation perform best. (Reproduced from ref. 38, with permission.)

Whereas the long hexagonally packed one-dimensional channels of MPS-5% and MPS-10% may induce some diffusional restrictions for Hg^{2+} to reach the binding sites located deep in the mesopores, transport within MPS-15% to MPS-30% sorbents is facilitated by a shorter range structural order in the form of three-dimensional wormhole framework structures. Increasing further the content of organic groups in the materials leads to poorly ordered (MPS-40% and MPS-50%) and even amorphous (MPS-70% and MPS-100%) solids, resulting in considerable reduction of mass transfer rates, to which the concomitant increase of hydrophobicity may also contribute to a significant extent. The differences in sorption rates exhibited by MPS materials appear therefore to result from differences in their relative long-range *versus* short-range structural order/disorder and their intrinsic hydrophobicity, which are induced by their functionalization levels.

1.8 Electrochemistry at the Sol–Gel Cage

Counter to intuition that would exclude non-conductive glassy materials from the field of electrochemistry, organically modified silica-based materials have a rich and varied electrochemistry[39] made possible by the accessible inner porosity. This allows oxidant and reducing reactant molecules to diffuse through the material and eventually to the surface either of a conducting electrode or of a conductive material (Figure 1.18).

The electrochemical deposition of doped silica, ORMOSIL and related nanocomposite thin films at the surface of electrodes is opening a whole new field of applications. In the latter case, a potential is applied to an alkoxide solution which alters the pH on the electrode surface and thus enhances the rate of condensation.[40] The resulting single-step

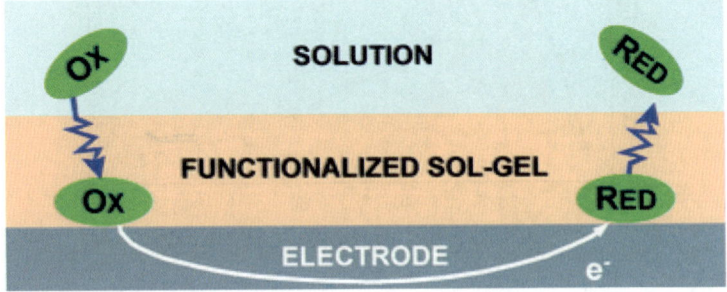

Figure 1.18   Electrochemical applications of ORMOSIL are due to the accessible porosity and the possibility to coat in thin films. (Reproduced from ref. 39, with permission.)

deposition is extremely fast and homogeneous, and the thickness and the porosity of the film is fully controllable by varying the potential and time of application. Applications in the fields of coating and catalysis followed soon after the discovery, and are discussed in the following chapters. Most recently, showing the potential of the methodology, a thin SiO_2–Cu nanocomposite coating was reported based on concomitant sol–gel deposition and reduction of copper ions.[41] Similar composites made of metal particles dispersed in dielectric materials are of great interest because they easily give rise to optical nonlinear phenomena such as super-magnetism.

Tunnelling atomic force microscopy (TUNA) of an electrochemically deposited Cu–SiO_2 film, providing simultaneously conductivity and topography images, clearly differentiates between the metallic conducting zones and the non-conducting sol–gel phase (Figure 1.19). From the overlap of the processed conductivity image with the topography image (Figure 1.19c), the higher topographic areas nicely correlate with enhanced conductivity, which implies that these areas are made of metallic copper.

Also for electrochemical applications, the possibility to fine tune the HLB provides the unique advantage of controlling diffusion in the material. Indeed, changing the HLB of the interfacial region of the host—the walls of the silica sol–gel cage—significantly alters the rate of diffusion of dopant molecules through the inner porosity of the matrix. For example, the addition of a small amount of –CH_3 groups to an electrode coated with doped sol–gel silica enhances the diffusion coefficient of entrapped $Co(bpy)_3^{2+}$ by a factor > 3.3.[42] The addition of a quaternary ammonium (using organoalkoxysilane containing such a functional

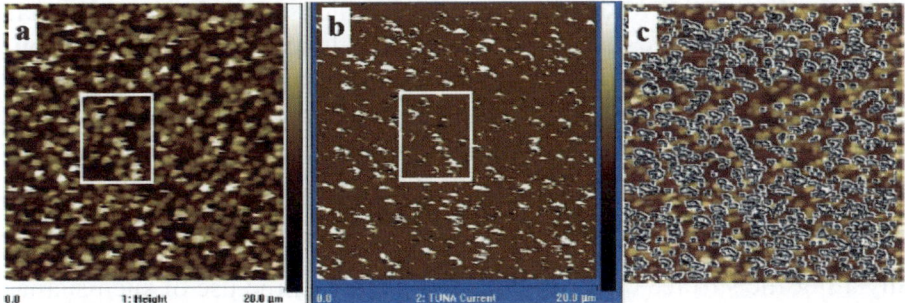

Figure 1.19   (a) Topography, (b) conductivity and (c) topography–conductivity overlap images obtained by TUNA of a Cu–SiO_2 film electrochemically deposited from 10 mM Cu^{2+} and TMOS upon applying −0.6 V for 10 min. (Reproduced from ref. 41, with permission.)

group) *enhances* the diffusion coefficient of the same cobalt species by one order of magnitude, and *reduces* that of $Fe(CN)_6^{3-}$ by one order of magnitude relative to values measured for SiO_2.[43] Reduction of activity with time is again due to the shrinkage of the xerogel porosity: a common problem with SiO_2-based gels that requires organic modification of the matrix and, in particular, optimal use of fluorinated organosilane as co-precursor of the ORMOSIL matrix.[44]

Encapsulation of microdroplets of the liquid crystal 4′-pentyl-4-biphenylcarbonitrile in ORMOSIL matrices prepared from TEOS and triethoxy derivatives results in materials with better transparency and thermal stability than polymer-dispersed liquid crystals (PDLCs) and with a larger refractive index differential between matrix and liquid crystal, leading to greater light scattering (opacity) in the OFF state (Figure 1.20).[45] Remarkably, samples prepared from TEOS or mixed TEOS–methyltriethoxysilane (Me-TES) precursors show switching times between ON and OFF states of 10 and 300 ms, respectively, and require peak-to-peak voltages (the driving voltage for operation) of 200–300 V. In contrast, samples using TEOS–Et-TES and TEOS–Pr-TES precursor mixtures require much lower activation voltages of 100 and 40 V, respectively. The procedure forms 2 mm thin gel–glass dispersed liquid crystal (GDCL) layers, a 10- to 25-fold decrease compared to previous state-of-the-art 20–50 mm layers, and showing similar electro-optical responses.

1.9   Symmetry: Chiral Objects and Chiral Cages

Chiral silicas are chiral objects first obtained by Moreau and co-workers in the early 2000s by transcription and self-assembly of bridged silsesquioxanes *via* cooperative weak interactions, affording chiral, left- and right-handed organosilica helices (Figure 1.17). On the other hand, shaping the cages (voids) of amorphous silica xerogels by molecular imprinting results in shape-selective materials ideally suited for applications such as separations, chemical sensing and catalysis. For bulk silica, the method was first demonstrated by aminopropyl-modified silica acting as shape-selective basic catalysts.[46] The concept is general, and is now widely employed for producing selective imprinted xerogel sensors ideally suited for bioanalytical applications.[47]

Synthetic methods include the use of silanes bearing a chiral group for silylating the surface of the porous sol–gel silica, the use of such silanes as monomers or co-monomers in the polycondensation, the physical entrapment of chiral molecules, the imprinting of SG materials with chiral templates and the creation of chiral pores, and the induction of chirality in the matrix skeleton itself.[48]

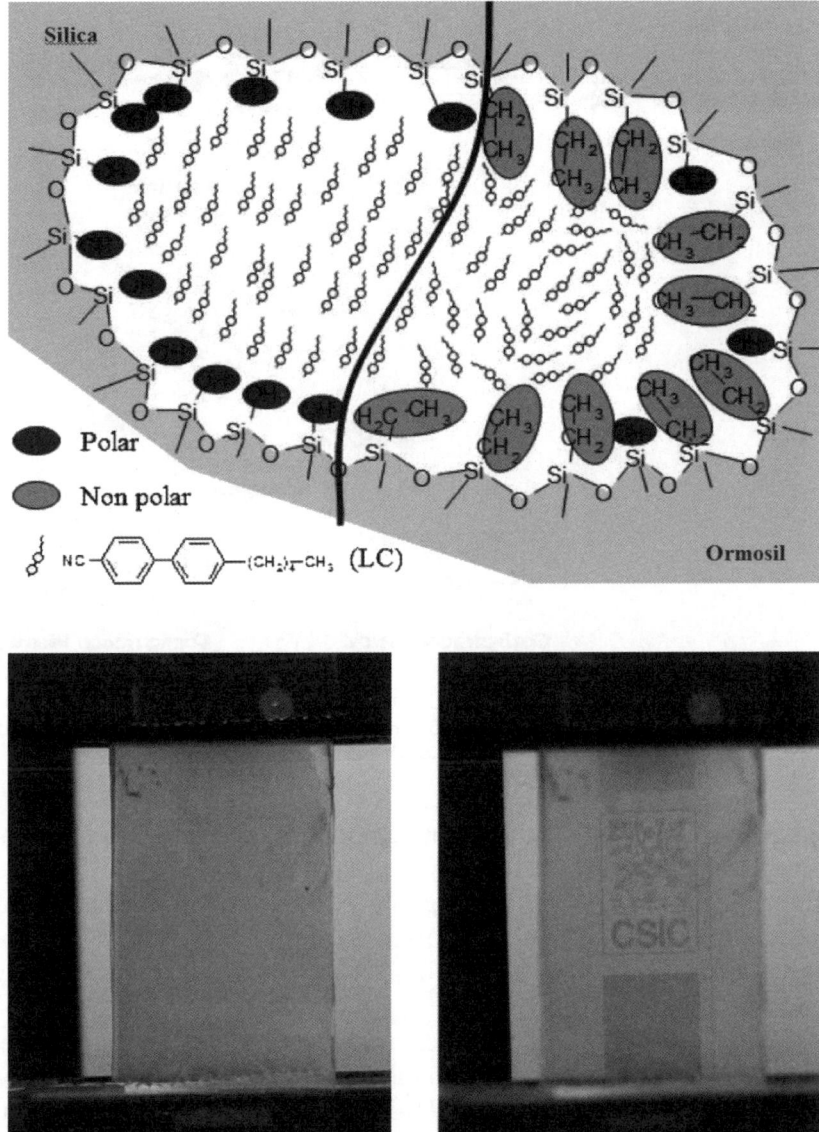

Figure 1.20   Encapsulation of microdroplets of liquid crystals in ORMOSIL matrices results in materials with better transparency and thermal stability than polymer-dispersed liquid crystals. Gel–glass dispersed liquid crystal device switched between the OFF and ON state (thickness 10 μm, 4 × 2 cm, $V_{p-p} = 90$ V). (Reproduced from ref. 45, with permission.)

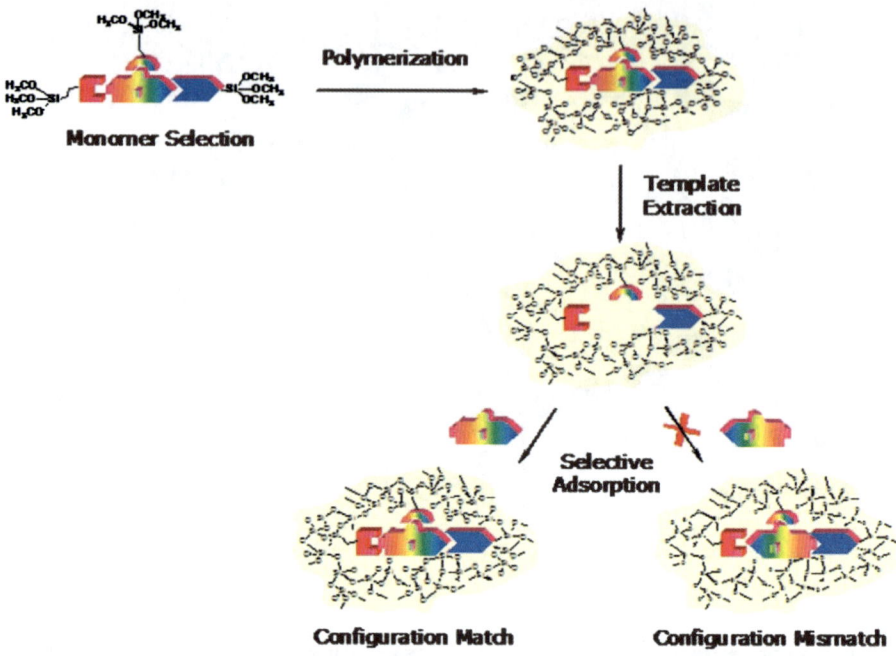

Figure 1.21 Schematic representation of the strategy of molecular chiral imprinting
of a sol–gel matrix using a chiral template and suitable achiral silanes.
(Reproduced from ref. 48, with permission.)

Molecular imprinting for inducing chirality in porous silica consists of
two stages: first, the three-dimensional structure is created during the
polymerization process by mixing the chiral template molecule with
alkoxysilane monomers; second, after the polymerization is complete,
the template is removed leaving behind its three-dimensional "negative"
chiral pore (Figure 1.21).

ORMOSIL are chemical sponges: they adsorb and *concentrate* reactants at their surface, thereby enhancing reaction rates and sensitivity
(in sensing applications). ORMOSIL-imprinted materials with a suitable
chiral "template" such as a surfactant or a protein selectively adsorb
(and detect) external reactants. A remarkable example is provided by
thin materials that are *generally* enantioselective, namely where the
chirally imprinted cavities can discriminate between enantiomers of
molecules not used in the imprinting process, and completely different
from the imprinting ones.

In this case, the material's specific chiral cavities are created by utilizing
phenyltrimethoxysilane and TMOS as the monomers and the chiral
cationic surfactant *N*-dodecyl-*N*-methylephedirnium (DMB) as template,

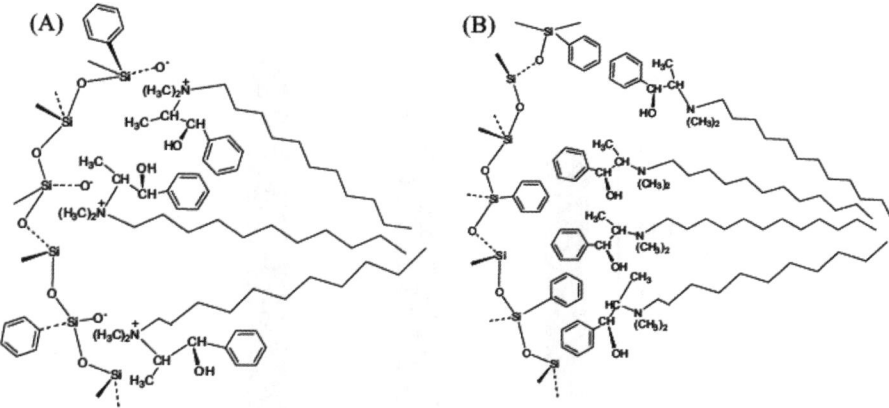

Figure 1.22   This surfactant orientation is supported by the *d*-spacing values that were obtained from transmission electron microscopy measurements. In (A) the first suggested orientation the surfactant length is 2.4 nm and in (B) another orientation the surfactant length is 1.85 nm. (Reproduced from ref. 49, with permission.)

followed by DMB removal with methanol, yielding a 20% phenyl-modified templated silica (DMB@PSG) monolith. The enantioselective discrimination here takes place either in water or in organic solvent and either through π–π interactions (between the long hydrophobic chains pointing away from the silica backbone and the formed matrix) or through electrostatic attraction in a folded manner (Figure 1.22).[49] The resulting imprinted silica, the chiral silicate polymer, is capable of discriminating between enantiomers of several molecules, none of which was used for the templating procedure (Figure 1.23).

1.10   Anisotropy and Dissymmetry: Silica Nanotubes and Janus Nanoparticles

Template synthesized silica nanotubes (SNTs) provide unique features such as end functionalization to control drug release, inner voids for loading biomolecules, and distinctive inner and outer surfaces that can be differentially functionalized for targeting and biocompatibility.[50] A general path to synthesize nanotubes utilizes anisotropic materials as template. They are coated with silica using $Si(OR)_4$ precursors and nanotubes of SiO_2 are obtained after removal of the template (Figure 1.24).

Diverse chiral nanometric ribbons and tubules formed by self-assembly of organic amphiphilic molecules can be transcribed to inorganic

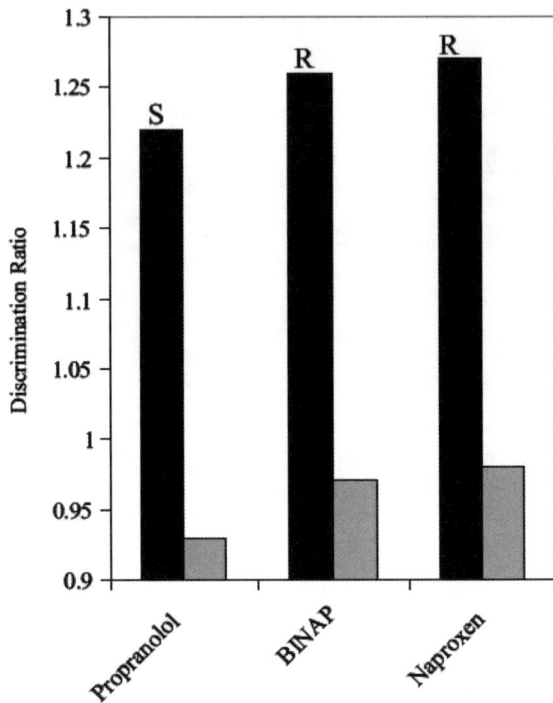

Figure 1.23 Discrimination ratios for the enantiomers propranolol, BINAP and Naproxen over DMB@PSG imprinted silica (black bars) and PSG reference material (grey bars). (Reproduced from ref. 49, with permission).

Figure 1.24 Silica nanotubes are formed by template synthesis using anisotropic materials ($V_3O7 \cdot H_2O$, in this case.)

nanostructures using a sol-gel transcription protocol with tetra-ethoxysilane (TEOS) in the absence of catalyst or cosolvent;[51] including open secondary architectures rather than a closed tube using ubules made of lipid surfactant ($DC_{8,9}PC$) that once added to water containing TEOS co-assemble with TEOS hydrolysing to produce silicate anions which interact with the lipid's cationic headgroups and deposit along the lipid bilayers as they twist into the helical shape (Figure 1.25).[52]

The dynamic and versatile nature of the organic gels considerably enhances the tunability of inorganic materials with rich polymorphisms (Figure 1.26), from nanotubes to nanohelices.

Hence, by controlling parameters such as temperature or the concentration of the different reactants, the morphology of the inorganic nanostructures formed from organic templates can be finely and widely tuned. Finally, long, double-walled glass nanotubes (fibres) ten times narrower than those currently available with a monodisperse diameter

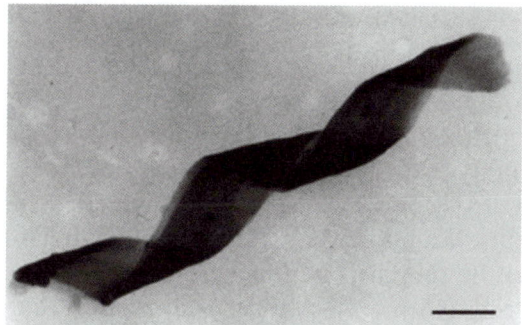

Figure 1.25 A silica–lipid mineralised helical ribbon (scale bar 400 nanometres). (Reproduced from ref. 52, with permission.)

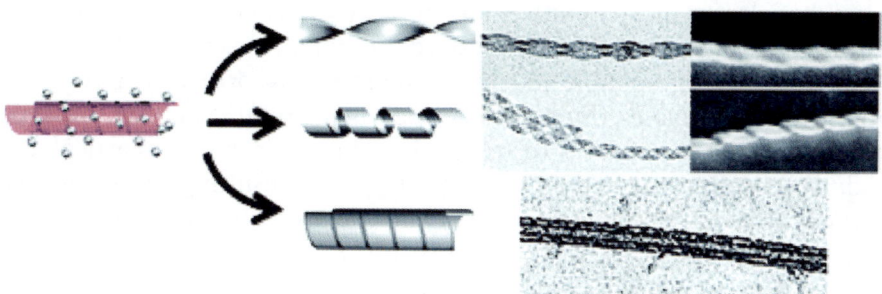

Figure 1.26 Template synthesis using organic gels expands the versatility of the methodology, affording largely different silica nanotubes. (Reproduced from ref. 51, with permission.)

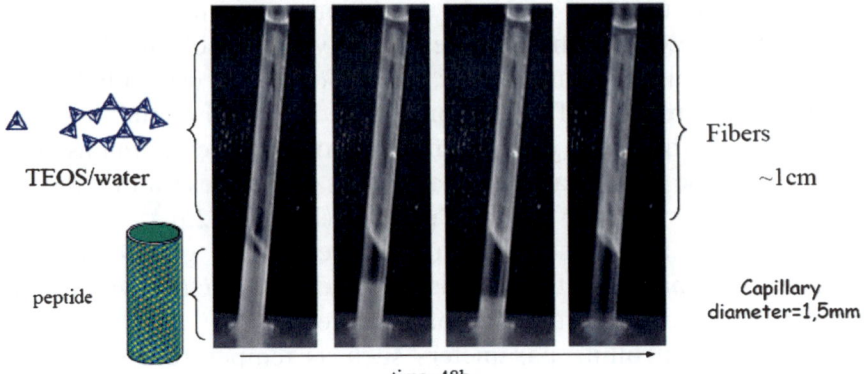

Figure 1.27   Double-walled silica nanotubes with monodisperse diameters self-orga-
nize into highly ordered centimetre-sized fibres, using a synthetic octa-
peptide as a template. The growth mechanism is proposed to be the
fundamental mechanism for growth processes in biological systems.
(Reproduced from ref. 53, with permission.)

can now be easily synthesized using as template synthetic peptide Lan-
reotide, that spontaneously self-assembles in water (5% w/w) forming
2 nm thick nanotubes with 24.4 nm diameter (Figure 1.27).[53]

Impressive, highly ordered centimetre-sized fibres are obtained whose
synergistic growth mechanism based on the kinetic cross-coupling of a
dynamical supramolecular self-assembly and a stabilizing silica minera-
lization may well be the basis of the synthetic paths used by Nature to
obtain its materials with well-defined multiscale architectures in biological systems.

Called "Janus" particles by de Gennes with reference to the double-
faced god of the Doors (Janus, in Latin), surface-dissymmetrical silica
particles were in the late 1980s. Today, relatively large amounts of Janus
nanoparticles (typically, one gram of Janus particles can be obtained
within two days) can be synthesized in batch based on the elegant
concept that a removable mask can temporarily protect a part of an
object in a reactive medium, while a polymer nodule is grown onto the
surface of silica particles to yield silica/polymer dissymmetrical col-
loids.[54] The unprotected mineral part of the resulting snowman-like
particles is selectively functionalized and the protecting polymer mask is
removed in a subsequent step (Figure 1.28).

The synthetic route for making the Janus nanoparticles first consists
in the emulsion polymerization of styrene in the presence of silica
nanoparticles surface-modified by polymerizable groups. Snowman-like
hybrid nanostructures are thus obtained with 85% yield in which the

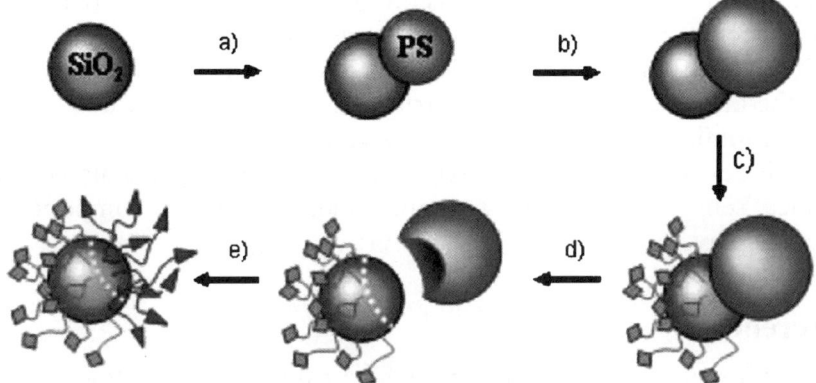

Figure 1.28 Schematic representation of the consecutive stages for the fabrication of silica Janus nanoparticles. (Reproduced from ref. 54, with permission.)

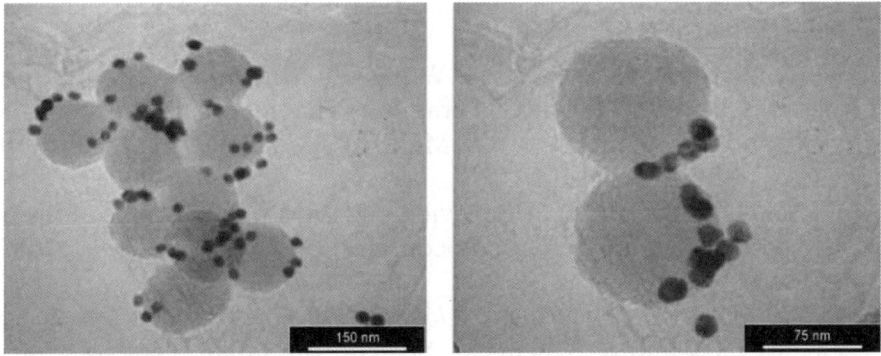

Figure 1.29 TEM images of Janus silica nanoparticles (diameter: 100 nm) in which the amine grafted area is decorated by gold nanocolloids, the remaining area being functionalized by methyl groups. (Reproduced from ref. 54, with permission.)

chemisorbed polymer chains generated at the surface of the silica particles forming one single nodule at the surface of the silica particle, as the high interfacial energy (due to the presence of unreacted silanol groups) does not promote their spreading on the surface.

The specific functionalization of the *unprotected* silica surface with a trialkoxysilane derivative such as $CH_3Si(OCH_3)_3$ is then carried out in a slightly basic water–ethanol suspension of the snowman-like particles. TEM images (Figure 1.29) of the resulting suspension of latex nanoparticles whose unmodified part was further functionalized with a second trialkoxysilane such as aminopropyltriethoxysilane and then treated with an aqueous suspension of citrate-stabilized 18 nm gold nanoparticles, clearly show the dissymmetrical Janus character of the resulting latex particles by specific adsorbtion of the metal nanoparticles onto the amine-grafted mineral surface. These nanostructures that can be amphiphilic, bifluorescent, responsive to an electric field (with both hemispheres of opposite charges), and will be useful as biological and chemical sensors, stabilizers of complex media, and nanocomponents in smart displays.

References

1. A. Fidalgo and L. M. Ilharco, Chemical Tailoring of Porous Silica Xerogels: Local Structure by Vibrational Spectroscopy, *Chem. Eur. J.*, 2004, **10**, 392.
2. D. Avnir, L. C. Klein, D. Levy, U. Schubert and A. B. Wojcik, *Organo-silica sol-gel materials*, in *The Chemistry of Organic Silicon Compounds*, ed. Z. Rappoport and Y. Apeloig, Wiley, Chichester, 1998, pp. 2317–2362.
3. M. Pagliaro, R. Ciriminna, M. Wong Chi Man and S. Campestrini, Better Chemistry through Ceramics: The Physical Bases of the Outstanding Chemistry of ORMOSIL, *J. Phys. Chem.*, 2006, **110**, 1976.
4. C. Sanchez, B. Julián, P. Belleville and M. Popall, Applications of hybrid organic–inorganic nanocomposites, *J. Mater. Chem.*, 2005, **15**, 3559.
5. D. Avnir, in *The History of Colloid and Surface Chemistry in Japan*, ed. K. Sakamoto, Chemical Society of Japan, Tokyo, 2001, pp. 16–17.
6. D. Avnir, D. Levy and R. Reisfeld, The nature of the silica cage as reflected by spectral changes and enhanced photostability of trapped Rhodamine 6G, *J. Phys. Chem.*, 1984, **88**, 5956.
7. D. Avnir, Organic Chemistry within Ceramic Matrixes: Doped Sol–Gel Materials, *Acc. Chem. Res.*, 1995, **28**, 328.
8. M. Pagliaro, R. Ciriminna and G. Palmisano, The chemical effects of molecular sol–gel entrapment, *Chem. Soc. Rev.*, 2007, **36**, 932.
9. A. Slama-Schwok, M. Ottolenghi and D. Avnir, Long-lived photoinduced charge separation in a redox system trapped in a sol-gel glass, *Nature*, 1992, **355**, 240.
10. M. L. Ferrer and F. del Monte, Study of the Adsorption Process of Sulforhodamine B on the Internal Surface of Porous Sol–Gel Silica Glasses through Fluorescence Means, *Langmuir*, 2003, **19**, 650.

11. P. Garcia Parejo, M. Zayat and D. Levy, Highly efficient UV-absorbing thin-film coatings for protection of organic materials against photodegradation, *J. Mater. Chem.*, 2006, **16**, 2165.
12. S. L. Murov, I. Carmichael and G. L. Hug, *Handbook of Photochemistry*, Marcel Dekker, New York, 2nd edn,1993.
13. H. Ow, D. R. Larson, M. Srivastava, B. A. Baird, W. W. Webb and U. Wiesner, Bright and Stable Core–Shell Fluorescent Silica Nanoparticles, *Nano Lett.*, 2005, **5**, 113.
14. H. Frenkel-Mullerad and D. Avnir, Sol–Gel Materials as Efficient Enzyme Protectors: Preserving the Activity of Phosphatases under Extreme pH Conditions, *J. Am. Chem. Soc.*, 2005, **127**, 8077.
15. A. Fidalgo, R. Ciriminna, L. M. Ilharco and M. Pagliaro, Role of the Alkyl–Alkoxide Precursor on the Structure and Catalytic Properties of Hybrid Sol–Gel Catalysts, *Chem. Mater.*, 2005, **17**, 6686.
16. R. Pardo, M. Zayat and D. Levy, The influence of sol-gel processing parameters on the photochromic spectral and dynamic behaviour of a naphthopyran dye in an ormosil coating, *J. Mater. Chem.*, 2005, **15**, 703.
17. R. S. H. Liu and G. S. Hammond, Reflection on Medium Effects on Photochemical Reactivity, *Acc. Chem. Res.*, 2005, **38**, 396.
18. R. Ciriminna, L. M. Ilharco, A. Fidalgo, S. Campestrini and M. Pagliaro, The structural origins of superior performance in sol–gel catalysts, *Soft Matter*, 2005, **1**, 231.
19. R. Abu-Reziq, D. Avnir and J. Blum, Three-Phase Microemulsion/Sol–Gel System for Aqueous Catalysis with Hydrophobic Chemicals, *Chem. Eur. J.*, 2004, **10**, 958.
20. G. Palmisano, R. Ciriminna and M. Pagliaro, Waste-Free Electrochemical Oxidation of Alcohols in Water, *Adv. Synth. Catal.*, 2006, **348**, 2033.
21. H. Frenkel-Mullerad and D. Avnir, The Chemical Reactivity of Sol–Gel Materials: Hydrobromination of Ormosils, *Chem. Mater.*, 2000, **12**, 3754.
22. P. W. Schmidt, D. Avnir, D. Levy, A. Hohr, M. Steiner and A. Roll, Small-angle x-ray scattering from the surfaces of reversed-phase silicas: Power-law scattering exponents of magnitudes greater than four, *J. Chem. Phys.*, 1991, **94**, 1474.
23. C. Rottman, G. Grader and D. Avnir D, Polarities of Sol–Gel-Derived Ormosils and of Their Interfaces with Solvents, *Chem. Mater.*, 2001, **13**, 3631.
24. D. A. Higgins, M. M. Collinson, G. Saroja and A. M. Bardo, Single-Molecule Spectroscopic Studies of Nanoscale Heterogeneity in Organically Modified Silicate Thin Films, *Chem. Mater.*, 2002, **14**, 3734.

25. N. J. Shirtcliffe, G. McHale, M. I. Newton, C. C. Perry and P. Roach, Porous materials show superhydrophobic to super-hydrophilic switching, *Chem. Commun.*, 2005, 3135.

26. C. Rottmann and D. Avnir, Getting a Library of Activities from a Single Compound: Tunability and Very Large Shifts in Acidity Constants Induced by Sol–Gel Entrapped Micelles, *J. Am. Chem. Soc.*, 2001, **123**, 5730.

27. M. J. van Bommel, T. N. M. Bernards and A. H. Boonstra, The influence of the addition of alkyl-substituted ethoxysilane on the hydrolysis—condensation process of TEOS, *J. Non-Cryst. Solids*, 1991, **128**, 231.

28. R. Ciriminna, S. Campestrini, M. Carraro and M. Pagliaro, Dynamic Catalysis in Aerobic Oxidation by Sol–Gel Living Materials, *Adv. Funct. Mater.*, 2005, **15**, 846.

29. W. J. Hunks and G. A. Ozin, Challenges and advances in the chemistry of periodic mesoporous organosilicas (PMOs), *J. Mater. Chem.*, 2005, **15**, 3716.

30. B. Alonso, C. Clinard, D. Durand, E. Véron and D. Massiot, New routes to mesoporous silica-based spheres with functionalised surfaces, *Chem. Commun.*, 2005, 1746.

31. A. Sayari and S. Hamoudi, Periodic Mesoporous Silica-Based Organic–Inorganic Nanocomposite Materials, *Chem. Mater.*, 2001, **13**, 3151.

32. B. D. Hatton, K. Landskron, W. Whitnall, D. D. Perovic and G. A. Ozin, Spin–Coated Periodic Mesoporous Organosilica Thin Films – Towards a New Generation of Low-Dielectric-Constant Materials, *Adv. Funct. Mater.*, 2005, **15**, 823.

33. C. Sanchez, H. Arribart and M. M. Giraud Guille, Biomimetism and bioinspiration as tools for the design of innovative materials and systems, *Nature Mater.*, 2005, **4**, 277.

34. R. Raja, J. M. Thomas, M. D. Jones, B. F. G. Johnson and D. E. W. Vaughan, Constraining Asymmetric Organometallic Catalysts within Mesoporous Supports Boosts Their Enantioselectivity, *J. Am. Chem. Soc.*, 2003, **125**, 14982.

35. R. Ciriminna and M. Pagliaro, Tailoring the Catalytic Performance of Sol-Gel-Encapsulated Tetra-n-propylammonium Perruthenate (TPAP) in Aerobic Oxidation of Alcohols, *Chem. Eur. J.*, 2003, **9**, 5067.

36. K. Kageyama, J.-i. Tamazawa and T. Aiola, Extrusion Polymerization: Catalyzed Synthesis of Crystalline Linear Polyethylene Nanofibers Within a Mesoporous Silica, *Science*, 1999, **285**, 2113.

37. L. Mercier and T. J. Pinnavaia, Access in mesoporous materials: Advantages of a uniform pore structure in the design of a heavy metal ion adsorbent for environmental remediation, *Adv. Mater.*, 1997, **9**, 500.

38. A. Walcarius and C. Delacôte, Rate of Access to the Binding Sites in Organically Modified Silicates. 3. Effect of Structure and Density of Functional Groups in Mesoporous Solids Obtained by the Co-Condensation Route, *Chem. Mater.*, 2003, **15**, 4181.

39. A. Walcarius, D. Mandler, J. A. Cox, M. M. Collinson and O. Lev, Exciting new directions in the intersection of functionalized sol–gel materials with electrochemistry, *J. Mater. Chem.*, 2005, **15**, 3663.

40. R. Shacham, D. Avnir and D. Mandler, Electrodeposition of Methylated Sol–Gel Films on Conducting Surfaces, *Adv. Mater.*, 1999, **11**, 384.

41. R. Toledano, R. Shacham, D. Avnir and D. Mandler, Electrochemical Co-deposition of Sol-Gel/Metal Thin Nanocomposite Films, *Chem. Mater.*, 2008, **20**, 4276.

42. M. M. Collinson and B. Novak, Diffusion and Reactivity of Ruthenium (II) Tris(bipyridine) and Cobalt (II) Tris(bipyridine) in Organically Modified Silicates, *J. Sol–Gel Sci. Technol.*, 2002, **23**, 215.

43. M. Kanungo and M. M Collinson, Controlling Diffusion in Sol–Gel Derived Monoliths, *Langmuir*, 2005, **21**, 827.

44. R. M. Bukowski, R. Ciriminna, M. Pagliaro and F. V. Bright, High-Performance Quenchometric Oxygen Sensors Based on Fluorinated Xerogels Doped with $[Ru(dpp)_3]^{2+}$, *Anal. Chem.*, 2005, **77**, 2670.

45. M. Zayat and D. Levy, The performance of hybrid organic–active-inorganic GDLC electrooptical devices, *J. Mater. Chem.*, 2005, **15**, 3769.

46. A. Katz and M. E. Davis, Molecular imprinting of bulk, microporous silica, *Nature*, 2000, **403**, 286.

47. E. L. Holthoff and F. V. Bright, Molecularly Imprinted Xerogels as Platforms for Sensing, *Acc. Chem. Res.*, 2007, **40**, 756.

48. S. Marx and D. Avnir, The Induction of Chirality in Sol–Gel Materials, *Acc. Chem. Res.*, 2007, **40**, 768.

49. S. Fireman-Shoresh, S. Marx and D. Avnir, Enantioselective Sol–Gel Materials Obtained by Either Doping or Imprinting with a Chiral Surfactant, *Adv. Mater.*, 2007, **19**, 2145.

50. Z. L. Wang, R. P. Gao, J. L. Gole and J. D. Stout, Silica Nanotubes and Nanofiber Arrays, *Adv. Mater.*, 2000, **12**, 1938.

51. T. Delclos, C. Aimé, E. Pouget, A. Brizard, I. Huc, M.-H. Delville and R. Oda, Individualized Silica Nanohelices and Nanotubes: Tuning Inorganic Nanostructures Using Lipidic Self-Assemblies, *Nano Lett.*, 2008, **8**, 1929.

52. A. M. Seddon, H. M. Patel, S. L. Burkett and S. Mann, Chiral Templating of Silica-Lipid Lamellar Mesophase with Helical Tubular Architecture, *Angew. Chem. Int. Ed.*, 2002, **41**, 2988.

53. E. Pouget, E. Dujardin, A. Cavalier, A. Moreac, C. Voléry, V. Marchi-Artzner, T. Weiss, A. Renault, M. Paternostre and F. Artzner, Hierarchical architectures by synergy between dynamical template self-assembly and biomineralization, *Nature Mater.*, 2007, **6**, 434.

54. A. Perro, S. Reculuse, F. Pereira, M.-H. Delville, C. Mingstaud, E. Dugnet, E. Bourgeat-Lamid and S. Ravaine, Towards large amounts of Janus nanoparticles through a protection–deprotection route, *Chem. Commun.*, 2005, 5542.

CHAPTER 2
Controlled Release

2.1 Sol–Gel Microencapsulation

Microencapsulation is a hot topic in the chemical industry and the subject of a growing number of scientific papers, reflecting research and development activities of chemical companies and research institutes. Technical papers are no longer published exclusively in the eight-issues-per-year *Journal of Microencapsulation* published since 1984, but in all leading international chemistry journals. Many microencapsulation companies (Figure 2.1) advertise on Google, and a simple web search using the "microencapsulation" query on that search engine yields some 234 000 results (as of May 2008).

There are many possibilities to consider when one needs controlled release of a chemical. Encapsulation can help eliminate a processing step, use less of an expensive ingredient or incorporate a new ingredient that would not work without encapsulation. Showing the great industrial relevance of encapsulation, Procter & Gamble, for instance, recently commercialized a microencapsulated liquid fabric softener[1] in the USA with the aim to keep its scent fresh through the cycle of washing, drying and wearing. Traditional low-volume markets for microcapsule-based products are expanding to include products such as adhesives, inks, fragrances, toners and sealants, and encapsulation companies now include chemical makers, flavour and fragrance houses, and specialist firms that offer competing technology.[2]

In a traditional microcapsulation process, the ingredient to be encapsulated is dispersed in a polymer-rich aqueous solution. The solvent characteristics of the medium are then changed, causing the

Silica-Based Materials for Advanced Chemical Applications
By Mario Pagliaro
© Mario Pagliaro 2009
Published by the Royal Society of Chemistry, www.rsc.org

 **Sol-Gel Technologies**

Figure 2.1   Companies such as Australia's Ceramisphere and Israel's Sol-Gel Technologies are changing the landscape of microencapsulation, commercializing the first sol–gel entrapped products.

Figure 2.2   The city of Beit Shemesh in Israel is the site of the world's first sol–gel microncapsulation company. During the hi-tech boom of 1996–2001 it was host to several hi-tech start-ups. (Photo courtesy of Wikipedia.)

polymer to undergo a phase separation. The polymer-rich phase spontaneously deposits itself as a coating around the dispersed ingredient and is solidified into a microcapsule, typically about 10 μm in size. The ingredient is later released when the microcapsule wall is ruptured by mechanical pressure or by friction.

Sol–gel microencapsulation in silica particles shares the versatility of the sol–gel molecular encapsulation process, with further unique advantages. Sol–gel controlled release formulations are often more stable, potent and tolerable than currently available formulations. The benefits of microencapsulation can be customized to deliver the maximum set of benefits for each active ingredient. Overall, these new and stable combinations of active pharmaceutical ingredients (APIs) result in improved efficacy and usability.

For example, Sol-Gel Technologies in Israel (Figure 2.2) manufactures on a large scale microcapsules made of sol–gel entrapped

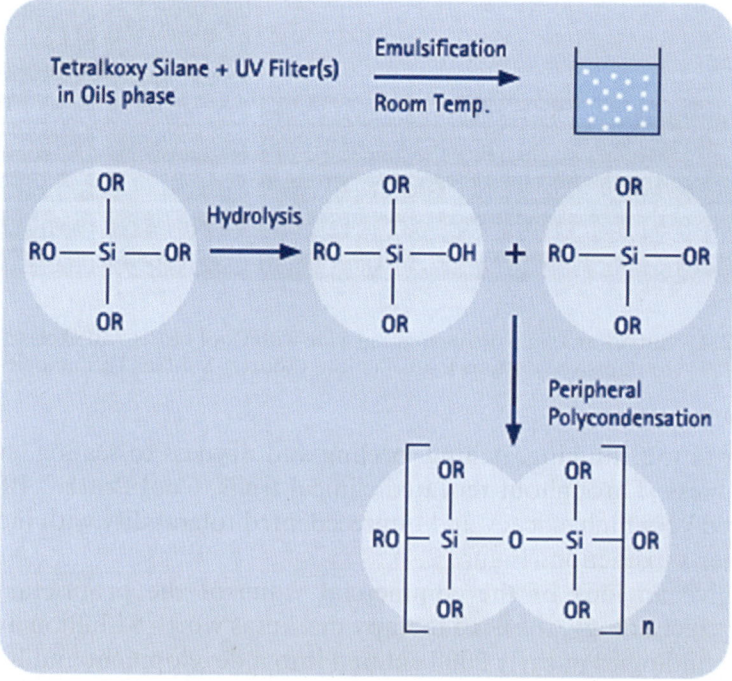

Figure 2.3 In the first step, the mixture is emulsified by stirring, and in the second step, an amorphous network of glassy material is prepared at room temperature by the hydrolysis of suitable monomers. The reaction proceeds to a condensation polymerization reaction, followed by subsequent formation of the sol to the gel and xerogel stages. (Adopted from Merck.com)

benzoyl peroxide (BPO) for effective, non-irritating treatment of acne. BPO crystals are encapsulated in transparent, porous silica shells by carrying out a sol–gel hydrolytic polycondensation in an emulsion phase (Figure 2.3).

The inert core shells serve as a safe protective barrier, preventing direct contact between the BPO and the skin and significantly reducing side effects. The product (Cool Pearls™ BPO) then precisely controls the amount of active ingredient deposited over time, adjusting it to meet the skin's individual needs. The mechanism involves migration of the skin's natural oily secretions through the silica pores into the capsule. The oils dissolve the BPO crystals and carry the BPO to the sebaceous follicles. The amount of skin lipids controls the rate at which the BPO is released, delivering optimal benefits. In brief, BPO entrapped in transparent silica pearls offers the proven benefits of BPO in anti-acne medication while virtually eliminating patient discomfort. Applied to the skin in simple formulations, BPO also produces numerous side effects ranging from

	Sol-Gel Technologies Cool Pearls BPO	Commercial Brand
The product significantly **cleaned my skin**	92%	40%
The product efficiently **cured my non-inflamed lesions**	92%	53%
The product produced **clean, soft skin**	67%	27%
The product produced **healthy looking skin**	75%	40%
The product **does not dry out the skin**	75%	8.3%

Figure 2.4   Survey among patients treating acne with Cool Pearls™ BPO clearly points to improved customer satisfaction. (Source: Sol-Gel Technologies Ltd.)

skin irritation, stinging, itching, peeling and dryness to scaling, swelling and redness. Throughout repeated clinical trials, Cool Pearls™ BPO has delivered very high efficacy and unprecedented tolerability with increased customer satisfaction (Figure 2.4).

To give an idea of the commercial value of the proprietary drug delivery technology, the acne therapy market is worth $1 billion and Sol-Gel Technologies in early 2008 entered into a development and licensing agreement with a US pharmaceutical company for which it will receive royalties from sales in the US and an extra $24.7 million to fund further product development aimed at other dermatology market segments in which BPO is widely employed.

Also, Sol-Gel Technologies manufactures inert glassy silica micro-particles encasing high concentrations of sunscreen molecules within a thin shell of inert sol–gel glass. With the increasing public awareness that ultraviolet (UV) sunlight is the primary cause of skin ageing, wrinkles and skin cancer, people use sunscreens more often, in higher concentrations (high protection factors) and in daily-use cosmetic preparations. A direct consequence of the increasing use of sunscreen molecules is that an increased amount of these molecules may penetrate through the epidermis into the body. Moreover, when UV light is absorbed by the sunscreen molecules, photodegradation products including free radicals may be formed and interact with body tissues. As UV filters can be encapsulated in glass microparticles (Figure 2.5), a protective and homogeneous UV-absorbing layer is placed on the skin's surface. The glass walls prevent interaction between the UV filters and skin. In other words, the encapsulated UV filters predominantly remain on the surface of the skin. Eusolex® UV-Pearls (Figure 2.5) are now used in the top formulations of the ever-increasing sunscreen market segment for sensitive skin, both young and aged.

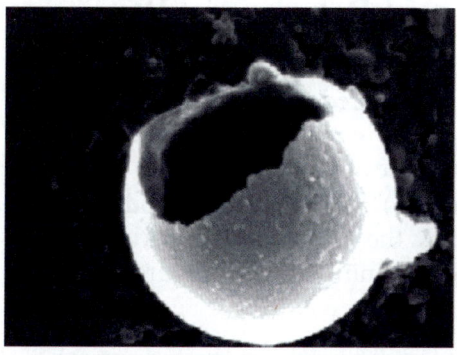 

Figure 2.5 A broken, spherical silica particle entrapping an API has 85% free volume. Such particles are used in formulations such as Eusolex® UV-Pearls™ that reduce dermal uptake compared to free UV filters; thus they do not irritate the skin while they make new application possibilities for hydrophobic UV filters. (Photo courtesy of Sol-Gel Technologies Ltd.)

The sol–gel microencapsulation here enables one to incorporate into a preparation a high ingredient load (*e.g.* 85% of the weight of the particles) and thus achieve sustained delivery of the active ingredients to the skin under defined mechanical or chemical conditions.[3] In addition, this technology affords low leaching (*i.e.* non-delivery) and enables the isolation of a component from its surrounding ingredients and the use of incompatible ingredients in the same formulation, because silica mineral and amorphous coating offers better tightness and resistance to extraction forces than polymers or waxes. The formulation of UV filters entrapped in silica microcapsules trademarked Eusolex® UV-Pearls™ is commercialized by Merck (with which Sol-Gel Technologies has engaged in an exclusive worldwide distribution agreement). The products are supplied as aqueous dispersions containing approximately 35% (w/w) of the UV absorber. The white liquids contain Eusolex® UV-Pearls™ of about 1.0 μm diameter on average. Ninety percent of the capsules are <2.5 μm in diameter, and are thus sufficiently small to be transparent when applied to the skin and to give a pleasant feeling. The dispersion consists of water, UV filter, silica, poly(vinyl pyrrolidone) and chlorphenesin, and shows excellent physical stability up to 80 °C and does not freeze. The pH is between 3.8 and 4.2. The aqueous microemulsion dispersion provides new opportunities for cosmetic formulators, as oil-soluble organic sunscreens can now be incorporated into the aqueous phase, and incompatibilities between cosmetic ingredients can be prevented to the benefit of novel combinations in a single cosmetic product.

2.2 Soluble Silica for Enhanced Release of APIs

An increasing number of new drugs developed consist of biopharmaceuticals, of which most cannot be administered orally and are chemically very unstable. Furthermore, there is an increasing need for non-oral advanced therapeutics delivery technologies that can maintain the viability of labile therapeutic agents (Figure 2.6). Sol–gel silica is biodegradable and is widely approved (for instance by the US Food and Drug Administration) for oral, mucosal and topical administration (Figure 2.7).

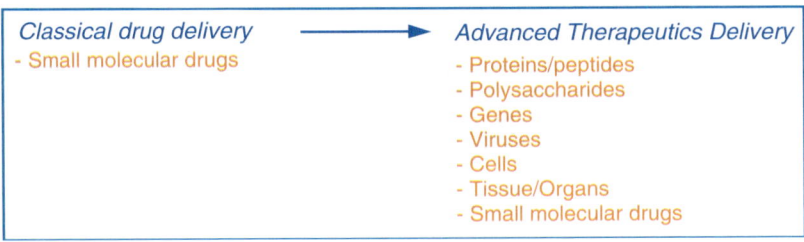

Figure 2.6   Trends in pharmaceutical technology.

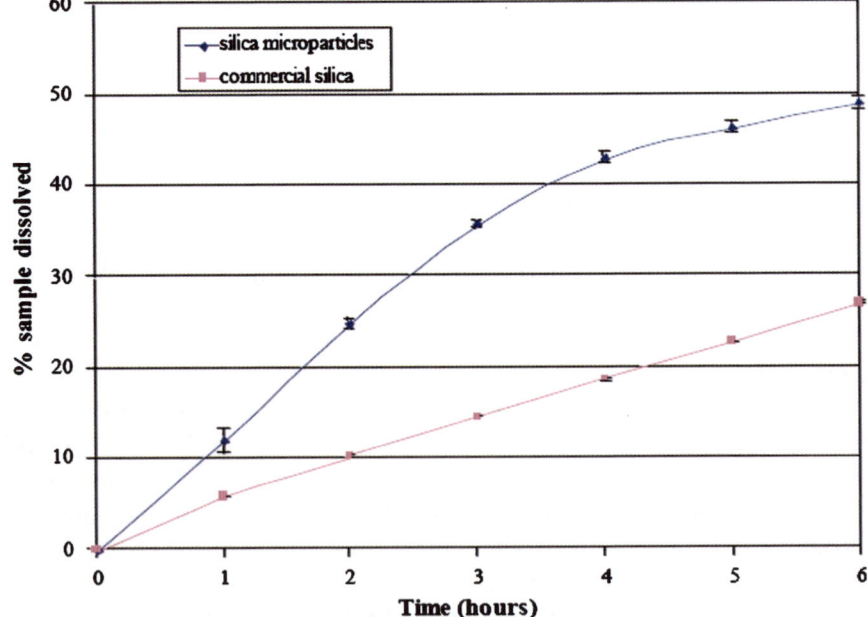

Figure 2.7   Sol–gel silica microparticles biodegrade faster than commercial silica.
(Reproduced from DelsiTech.com, with permission.)

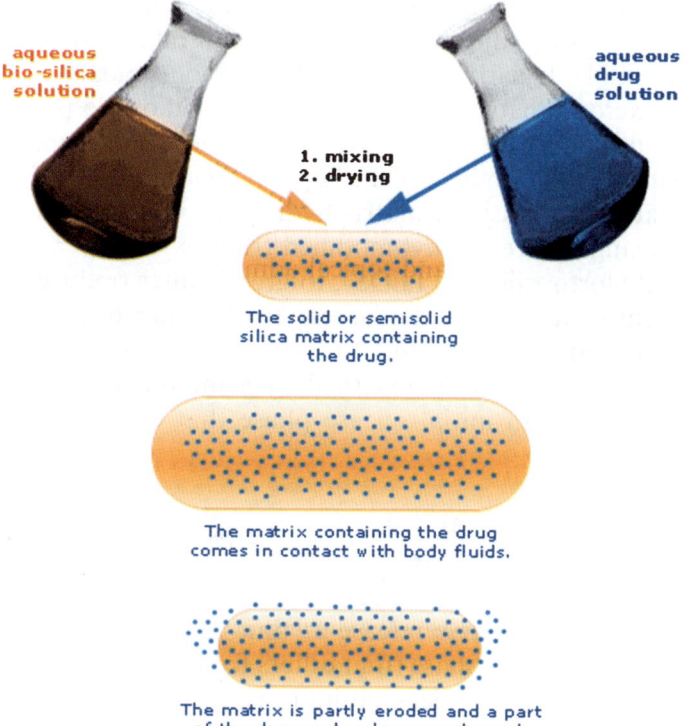

Figure 2.8   In DelSiTech technology, the API is entrapped into a SiO₂ matrix followed by dissolution and release on contact with body fluids. (Reproduced
from DelsiTech.com, with permission.)

   In one approach for the controlled delivery of a wide range of active
ingredients, including viruses, pursued by Finland's start-up company
DelSiTech, the technology is based on a SiO_2 matrix into which the
active ingredient is first embedded followed by dissolution and release
on contact with body fluids (Figure 2.8).
   The silica matrix can be designed to degrade at the required rate to
ensure a tightly controlled release of the active substance, providing a
better therapeutic effect *in situ* and causing lower systemic toxicological
side effects than alternative delivery systems. The proof of principle has
been demonstrated in animal testing both with small synthetic drugs and
biologics such as viruses. The advantages of this drug delivery technology
are evident in terms of safety (silica is a biocompatible natural component
of the body, does not affect the pH of the surrounding tissues, reducing
risk of inflammation) and enhanced performance (encapsulation is suitable
for synthetic small molecular compounds and for biopharmaceuticals; the

resulting solid matrix biodegrades, can be shaped in variable forms and delivery times can be adjusted from days to several months). As a representative example, heparin incorporated into biodegradable, sol–gel produced silica xerogel matrix is subsequently released over a prolonged time period retaining its biological activity, with the degradation of xerogel being linear and independent of the heparin concentration (Figure 2.9).

It is possible to modify the matrix hydrophilicity–lipophilicity balance (HLB) by using different alkyl-modified trialkoxysilanes as co-reagents with tetraethylorthosilicate (TEOS), to greatly alter (reduce) the release rate of heparin, as then water access to the entrapped drug is considerably reduced (Figure 2.10).[4]

The release mechanism is governed by a modified power law which takes into account an initial release phase that differs from the main release phase:[5]

$$\frac{M_t - M_2}{M_\infty} = k(t - t_2)^{n_B} \tag{2.1}$$

where M_t is the cumulative release at time t, M_2 the amount released within the initial release phase (that differs from the main release phase), M_∞ the cumulative drug release at infinite time, k a constant depending on the structural and geometrical characteristics of the device, t_2 the end point time of the initial release and n_B the release exponent

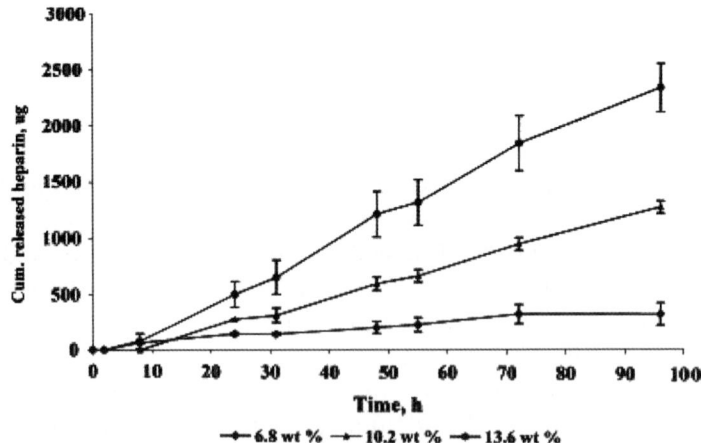

Figure 2.9   Cumulative release of heparin from silica xerogel matrix with varying drug concentration. Each data point represents the mean of three samples and error bars are standard deviations. (Reproduced from ref. 4, with permission.)

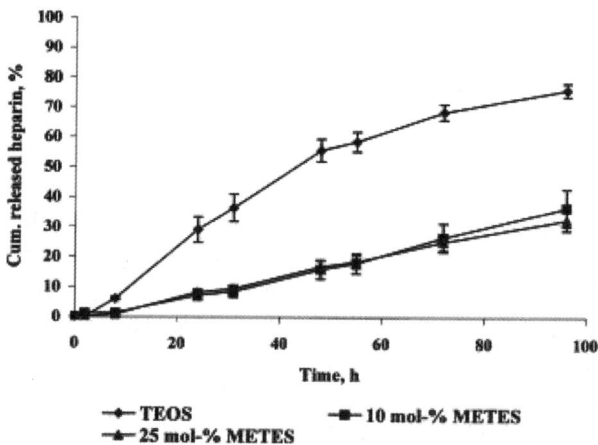

Figure 2.10 Cumulative release of heparin from methyltriethoxysilane-modified silica xerogel matrix. Each data point is the mean of three samples and error bars are standard deviations. (Reproduced from ref. 4, with permission.)

characteristics for the mechanism of drug release after the initial release. By plotting $\ln(M_t - M_2/M_\infty)$ *versus* $\ln(t - t_2)$, n_B can be calculated from the slope of the curve. The values of n_B describing release after the initial release are characteristic for the drug release mechanism.

In brief, diffusion is the main mechanism controlling the release of small-sized molecules from the SiO_2 monolith matrix with dissolution of the SiO_2 matrix having a clear effect on the release of monoliths made at low H_2O:TEOS ratios. Release of large molecules such as proteins prepared from high H_2O:TEOS ratio precursors occurs predominantly by matrix erosion.

The main use of the technology is for implants, subcutaneous and intramuscular administration mainly for cancer, neurological and orthopaedic applications. The technology is protected by nine international patents,[6] and recently the company entered into a licence agreement with Bayer, giving the large drugs manufacturer the exclusive rights to use the delivery technology for fibre applications for wound care.

2.3 Silica Microparticles for Controlled Release

Using sol–gel technology combined with water-in-oil (W/O) emulsions, a number of silica-based ceramic particles with independent control over the release rate and particle size are to be commercialized by Australian company CeramiSphere (and other companies) for a range of applications such as drug delivery, release of speciality chemicals and cosmeceuticals/nutraceuticals.

The average particle size can be varied from 10 nm to 100 μm and is controlled by the emulsion chemistry (Figure 2.11). The release rate can be independently varied from milligrams per hour to milligrams per month, and is controlled by the internal microstructure of the particles and the initial sol–gel chemistry.[7] Sol–gel emulsions are more effective because they enable the production of either micro- or nanoparticles with *homogeneous* distribution activity, and permit ambient temperature processing, which is necessary for handling temperature-sensitive biological agents. Independent control of the size and release rate can be readily achieved, and emulsion polymerization is easily scalable due to the compartmentalization of the reactions inside the emulsion droplets (Figure 2.12).[8]

Control of the particle size while retaining precise control over the release rate is enabled by compartmentalization of the sol–gel solution into droplets of definite size. This can be achieved by emulsification of the sol–gel solution by mixing it with a solution composed of a surfactant and a non-polar solvent (Figure 2.13). When an active molecule is located in the aqueous droplet of a W/O emulsion, encapsulation occurs as the silicon precursors polymerize to build an oxide cage around the active species. By changing the solvent–surfactant combination, the particle size can be varied from 10 nm to 100 μm as the size of the particles is controlled by the size of the emulsion droplet, which acts as a nano-reactor for the sol–gel reaction (Figure 2.13).

Upon mixing, the polar sol–gel droplets are dispersed in the non-polar solvent and act as micro-reactors in which the gelation proceeds, yielding microparticles with size comparable to that of the droplets. Thus, one can control independently the particle size by controlling the

Figure 2.11   Tailored particle size is ensured by the emulsion chemistry, as the droplet size where the sol–gel polycondensation takes place is easily controlled by the emulsion parameters. (Reproduced from ref. 8, with permission.)

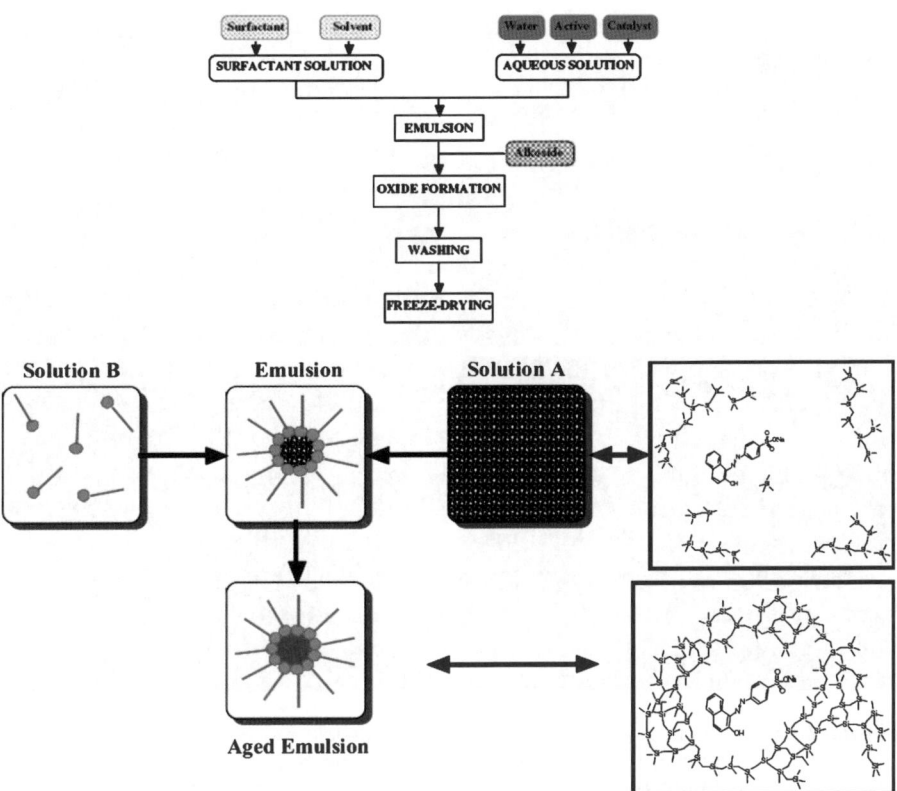

Figure 2.12 Nanoparticle synthesis flow-chart (top) and schematic of the combined microemulsion/sol–gel process to produce doped silica microparticles. (Reproduced from ref. 7, with permission.)

droplet size through the emulsion parameters and the release rate by controlling the internal structure of the particles through the sol–gel processing parameters. In particular, the release profiles can be tailored by controlling the internal structure of the particles: pore volume, pore size, tortuosity and surface HLB. This can be easily achieved by controlling sol–gel processing parameters such as the water-to-alkoxide ratio, pH, alkoxide concentration, ageing, drying time and temperature (Figure 2.14).

CeramiSphere technology is not limited to the encapsulation of small water-soluble molecules. It is also used to encapsulate hydrophobic molecules such as essential oils, flavours, vitamins, proteins (including enzymes) and many other biomolecules (such as DNA).

Biocompatible silica nanoparticles prepared by analogous sol formation followed by emulsification and doped with the antibiotic

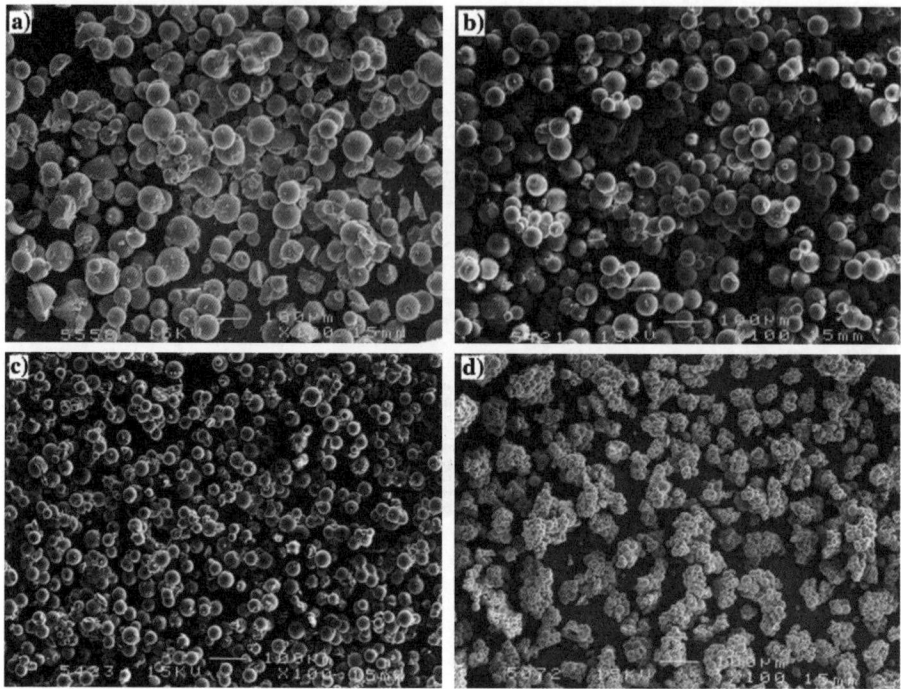

Figure 2.13 Influence of emulsion solvent on particle morphology: (a) hexane, (b)
octane, (c) dodecane and (d) cyclohexane. Surfactant is Span 80. All scale
bars are 100 μm. (Reproduced from ref. 8, with permission.)

vancomycin, for example, can thus be used for the treatment of bone
infections.[9] A xerogel doped with vancomycin, for example, shows a long-
term sustained release (up to six weeks). The kinetics of release and the
amount released are dose dependent. The initial, first-order release is
followed by a near-zero-order release. Regardless of the load, about 70%
of the original vancomycin content is released by the transitional point,
and the cumulative release after six weeks of immersion is about 90%.
Sol–gel microspheres, in particular, allow controlled, load-dependent and
time-dependent release. Variations of the water:alkoxysilane molar ratios
from 2 to 10 greatly affect the xerogel properties, and, consequently, the
release characteristics, with release from high molar ratio xerogels being
faster than release from low molar ratio xerogels due to larger surface
area and pore volume in xerogels obtained with higher water:alkoxysilane
molar ratio (Figure 2.15).

In comparison to a fast, short-term release from sol–gel granules, the
release from the microspheres is slower and of longer duration. In
addition, the degradation rate of microspheres is significantly slower

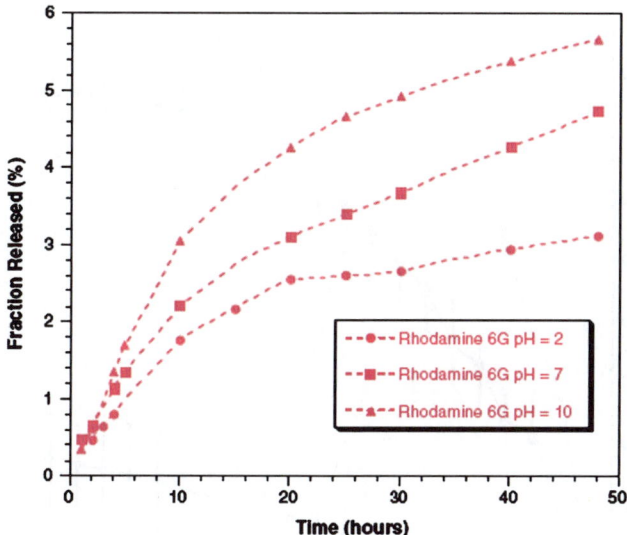

Figure 2.14 Influence of the synthesis pH on the release rate of Rhodamine 6G from microparticles. (Reproduced from ref. 7, with permission.)

than that of granules enabling the use of sol–gel powders for controlled long-term release.[10]

In general, sol–gel encapsulation ensures chemical protection of the valued entrapped dopants. Encapsulation provides

- prevention of degradation of proteins during passage through the gastrointestinal tract (acid + pepsin);
- prevention of enzymatic degradation of DNA and RNA;
- protection of enzymatic degradation of growth factor in wounds;
- protection from oxidation of small molecules such as retinol or flavours;
- protection from extreme acid or base; and
- protection from detergents.

This opens up applications in at least three major business sectors that will be profoundly impacted by the technology:

- *Healthcare*
  - Gene therapy
  - Wound healing
  - Oral delivery of protein
  - Vaccine (veterinary)

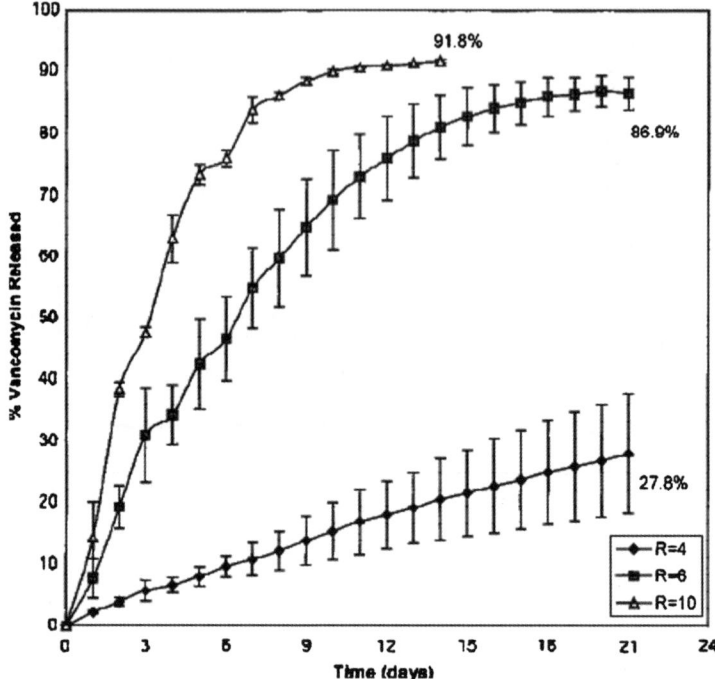

Figure 2.15   Percentage of incorporated vancomycin released from discs of load
20 mg g^{-1} as a function of time and water:tetramethylorthosilicate
(TMOS) molar ratio (*R*) (*n* = 3, errors bars represent ± 1 standard
deviation). Average total vancomycin amount in discs of water:TMOS
molar ratio = 4 of load 20 mg g^{-1} was 8.5 mg per disc. (Reproduced from
ref. 10, with permission.)

- *Speciality chemicals*
  - Biocide for building applications
  - Anticorrosion
  - Enzyme in washing powders
- *Cosmetics and foods*
  - Encapsulation of cosmeceuticals
  - Encapsulation of flavours and other food additives.

2.4   Therapeutic Films

Surface coating of mesoporous silicas provides an effective and flexible
way for controlled release due to the high drug loading possible and
the easy control of release rate. A drug-containing core or tablet is

surrounded by a sol–gel coated film, and the release rate of the drug is controlled by its diffusion through the film. In comparison with polymer and commercially available pharmaceutical tablet coatings, silica-based sol–gel coatings possess several advantages. Drug release can be easily controlled by changing the organosilane type and number of coatings. The precursor sols are of low viscosity, allowing several conventional coating methods. The textural and chemical properties of the final gel coatings can be adjusted by using different organosilanes and changing the sol composition. This is clearly shown, for example, by coating a mesoporous silica xerogel used as a carrier material for water-soluble vitamin B1 (VB1) release.[11]

The mesoporous silica carrier ensures a high drug loading, while the surface coating provides an easy, flexible and effective strategy for the controlled release. Design of a suitable carrier material for the controlled release is important because VB1 is not accumulated within the body and the risk of hypervitaminosis is practically absent. To investigate the effects of surface sol–gel coating on the release, a series of organosilanes can be used to coat VB1-loaded silica gel tablets.

The effects of organosilane type on the VB1 release are large and show that for tablets coated twice and dried at 60 °C, the samples coated with bis(trimethoxysilyl)hexane (TSH), bis(triethoxysilyl)octane (TSO), bis (trimethoxysilylpropyl)amine (TSPA) and bis[3-(trimethoxysilyl)propyl]-ethylenediamine (enTMOS) give slower release than the other samples including uncoated mesoporous silica used for a comparison (Figure 2.16).

There remains much need for materials useful in surgery, in therapeutics, for the treatment of wounds and other applications that allow the controlled release of pharmaceutically active molecules. It has long been desired to realize materials that, for example, are bacteriostatic and can be used in emergent therapy for wounds. Other materials are desired for use in surgery, especially orthopaedic surgery, while still other uses involving such controlled release of medicaments will find immediate application in diverse therapeutic areas. Xerogel films on substrates are available containing pharmaceutically active compounds. Materials incorporating such films are robust, release active compounds at predictable rates and may provide such release for relatively long periods of time. Orthopaedic and trauma uses are indicated along with generalized use in contact with body fluids or as biological implants.[12]

The coatings clearly show a time- and load-dependent release. The rate of release and the total amount released from coatings with 20% vancomycin are significantly greater than those from coatings with 10% vancomycin (Figure 2.17). At both 10 and 20% concentration, the initial faster release with subsequent slower release was observed. For both

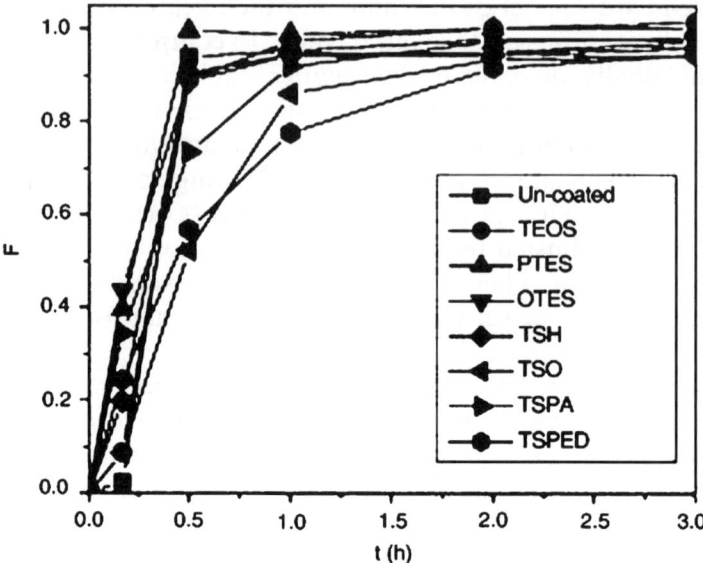

Figure 2.16 Effect of sol–gel coating precursors on release. Tablets were coated twice and dried at 60 °C. The release media were 0.05 M phosphate buffer solutions at pH = 7.4. (Reproduced from ref. 11, with permission.)

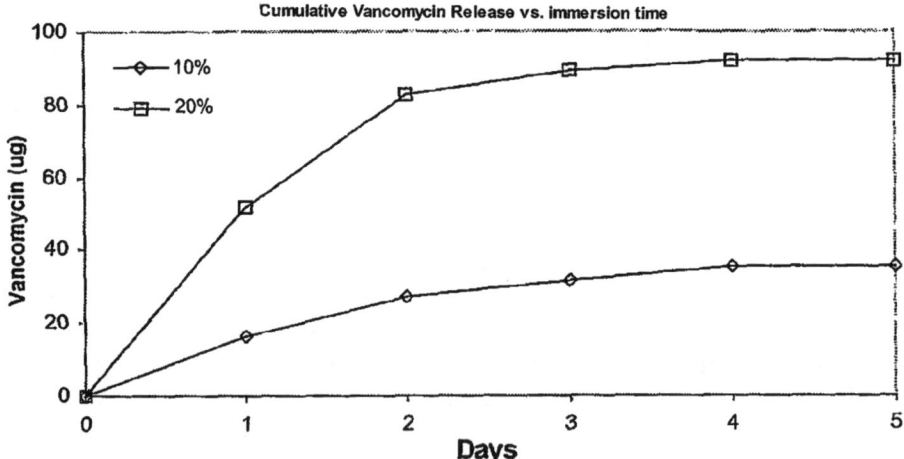

Figure 2.17 Cumulative vancomycin release of single-layer coatings as a function of antibiotic concentration. (Reproduced from ref. 12, with permission.)

concentrations, about 80% of the original vancomycin load was released after two days of immersion.

Even very thin films of xerogels can release useful amounts of pharmaceutically active material into body fluids and the release kinetics are stable and predictable. Concomitantly, a relatively long release profile, of the order of weeks or months, may be attained, making these materials flexible and useful for long-term therapy. Xerogel films that have excellent mechanical adhesion to substrates (evaluated via press-fit test), especially to metal substrates, are very useful for implantation in patients. For example, adhesion tests of vancomycin-loaded thin xerogel film to ano-dized titanium wires did not show any evidence of cracking in implanted samples. Coating was as simple as putting 30 mm long wires in contact with a TEOS-derived sol. Typically, the films were composed of five layers applied on the wire surface by dip coating the wire with vancomycin-doped sols at a withdrawal speed of $80\,\mathrm{mm\,min^{-1}}$. Extending this approach, it has been shown that large amounts (up to 20%) of the water-soluble antibiotic vancomycin can be incorporated into a sol–gel film deposited on titanium alloy substrate.[13] The release properties and the film stability largely depend on the number of applied layers and the van-comycin load. The release rates and the total amount released increase with the number of layers and the load. In addition, the time required for total release also increases with the number of layers. Optimized

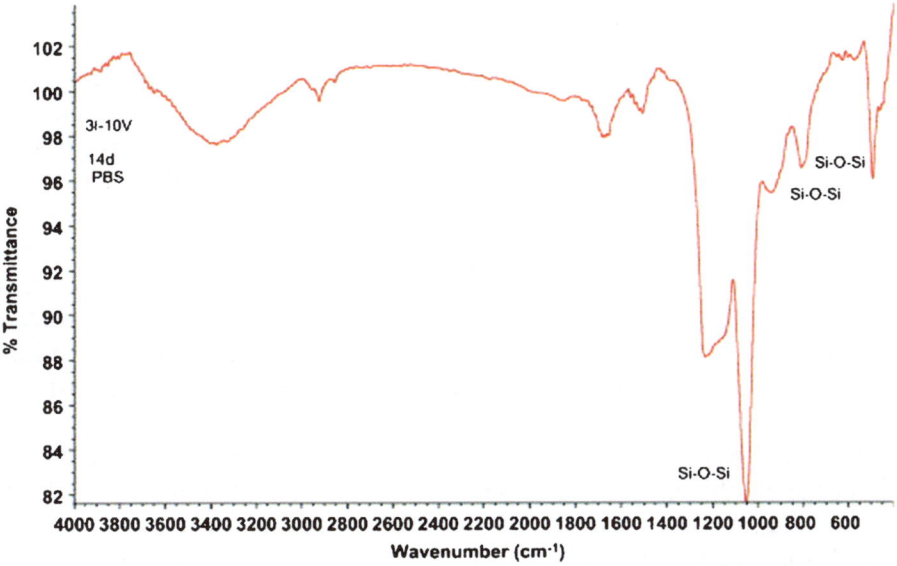

Figure 2.18 Fourier transform infrared spectrum of sol–gel film composed of three layers and containing 10% vancomycin after 14 days of immersion in phosphate-buffered saline (weight measurements and release study showed 90% weight loss and 80% release). Intense silica bands and the bands associated with vancomycin (centred at 1660, 1500 and $1397\,\mathrm{cm^{-1}}$) suggest the presence of a sol–gel film with vancomycin that remains after 14 days of immersion. (Reproduced from ref. 13, with permission.)

multilayer films showed a zero-order release of vancomycin up to two weeks. Moreover, Fourier transform infrared (FTIR) analysis reveals that a thin remaining film with vancomycin was still present on the titanium alloy surface after the two-week immersion period (Figure 2.18).

There is a difference between the release mechanisms from thin sol–gel films and from bulk xerogels. This difference in release mechanisms is probably related to nanostructural differences between monolithic sol–gels and thin sol–gel films. Monolithic xerogels are produced via casting, gelation and drying, and are highly porous materials with a large surface area and a large pore volume. Solution penetration into the porous matrix allows biomolecules to diffuse out. Unlike monolithic sol–gels, for thin sol–gel films all stages of sol–gel transformation overlap and occur at once during the film deposition step. Thin sol–gel films derived from acid-catalysed, unaged sols may have a very low porosity to the point of being nonporous unaged), they are largely nonporous. Although, in principle, solution penetration into such films is possible, it is also limited.

2.5   Controlled Release of Enzymes

Laundry detergents today include enzymes (often bioengineered ones) that break down proteins, starches, cellulose and other components of fabric stains. Biological washing powders usually contain enzymes to help digest stains while the latest shampoos and hair conditioners often use proteins to add shine and thickness to hair. These enzymes, however, must be kept stable against mutual destruction by other protein-destroying enzymes and against other chemicals in the product. Hence it is necessary to protect stain-fighting enzymes during storage; however, the same protected enzymes must be readily available when the product is put to use. Microscopic gel beads made of sol–gel organosilica-entrapped enzymes meet the requirements.[14]

These gel beads are synthesized from the precursor bis[3-(trimethoxy-silyl)propyl]ethylenediamine (enTMOS) which undergoes sol–gel processing in the presence of dissolved alkaline proteases such as subtilisin. Encapsulated enzymes are found to be stable in these gels for extended periods of the order of years. Yet, the gels thereby formed are aggregations of loosely interacting particles, which can be easily dissolved when placed in water, leading to a fast release of dopant entities. Figure 2.5 shows the highly porous structure formed by aggregation of particles. In concentrated form, the gel beads remain concentrated and tightly sealed. However, when water is added the

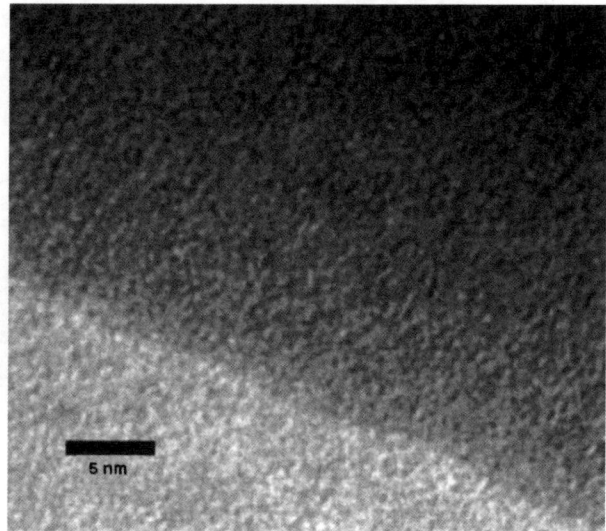

Figure 2.19 Transmission electron micrograph of a K3 sol–gel sample. (Reproduced from ref. 14, with permission.)

concentration drops and the beads are free to expand. It is this swelling process that is key to protecting the biomolecules until they are needed.

The dopant molecules including the enzyme are trapped in this network without significant strain on the native conformation, thereby retaining the activity of the enzyme (Figure 2.19). FTIR data indeed show interactions of the formulated enzyme during sol–gel matrix formation such that the final material has reduced hydrogen-bonding interactions and as a result forms a less cross-linked structure. Through a judicious choice of precursor composition, it is possible to optimize the composition for effective stabilization of the enzyme as well as efficient release when dissolved.

Initially, the sol–gel compositions were optimized using Congo red dye as the dopant because of its optical properties. This facilitates monitoring of the release process by optical spectroscopy. Next, the gels were evaluated for their stabilization and release of subtilisin. These sol–gel matrices bring about controlled release of the encapsulated enzyme molecules as a response to a change in the water content of the medium (Figure 2.20).[15]

The enzymes are stabilized due to hydrogen-bonding and electrostatic interactions with –NH and SiOH groups. Furthermore, the gels retain a significant amount of water necessary for the stability of the encapsulated biomolecules. As a result, almost all of the enzyme (K1, 94%; K3, 98%; K4, 100%; K5, 100%) is stable and is released from the gels when placed

Figure 2.20   Dissolution-controlled release from organosilica sol–gels: a graphical representation of the process of dissolution of sol–gels that accompanies release of encapsulated molecules along with a schematic representation of changes in absorbance of the solution as a function of time that can be used to monitor the release process. (Reproduced from ref. 15, with permission.)

in water. The use of a modified sol–gel protocol whereby the sols were aged for a period of 30–90 min prior to the addition of the enzyme leads to substantial enhancement in retention of native activity (Figure 2.21).

Furthermore, there is minimal loss of activity, even after storage of the sol–gels with enzyme under ambient conditions for an extended period of time. Clearly, long-term storage stability is a critical requirement in a fabric care system. Here the sol–gel encapsulation ensures such long-term stability of the enzymes in the gels stored under ambient conditions, avoiding denaturation of the enzymes after ageing of the sol–gels (ageing usually causes the silicate network to become more extensive and rigid, and the local environment more conducive to protein–silica interactions with potential protein unfolding due to electrostatic interaction of the active sites of the enzyme with the silicate matrix).

The swollen gels were tested as an additive to two different detergents: the first was a generic detergent ingredient, concentrated sodium dodecyl sulfate (SDS); the second was a commercial laundry product with its enzyme content destroyed by heating to 90 °C for an hour. The gels remained tight and contracted in the detergents, but as water was added, they began to swell and release their contents into the wash. Furthermore, the enzyme in the swollen gels is much more stable and able to function even after heating to 80 °C as opposed to total loss of activity with a control experiment using a standard enzyme solution.

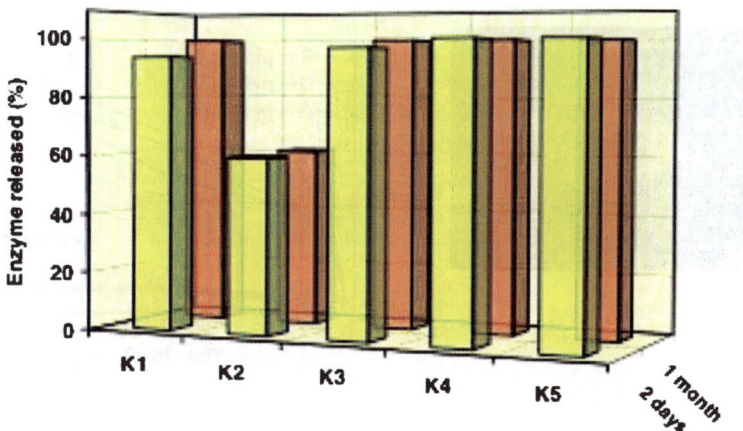

Figure 2.21 Percentage of subtilisin released upon dissolution of sol–gels aged for 2 days and for 1 month. All the sol–gels except K2 were prepared by adding enzyme to a pre-hydrolysed aged sol. (Reproduced from ref. 15, with permission.)

2.6 Controlled DNA Delivery

Another major example of the active role played by the sol–gel cage HLB in dictating reactivity and accessibility of the entrapped dopant concerns the first case of genes delivery into the (mouse) brain. Indeed, it has been recently demonstrated for the first time that organosilica nanoparticles are promising candidates for efficient DNA delivery in the brain.[16] The transfection efficiency of organically modified silicates (ORMOSIL) was equal to or even better than that of herpes simplex virus-1 (HSV-1). Moreover, ORMOSIL-mediated delivery does not cause the tissue damage or immunological side effects that have been commonly observed with viral-mediated gene delivery. Hence, in place of dangerous viruses traditionally used as gene delivery vehicles, ORMOSIL amino-functionalized nanoparticles obtained by adding 3-aminopropyltriethoxysilane to a polymeric precursor made of tri-ethoxyvinylsilica chains self-assembled from $(H_2C=CH)Si(OEt)_3$ in the core of an oil-in-water microemulsion (surfactant (aerosol-OT)–dimethylsulfoxide–water) can efficiently introduce a healthy version of a gene into a patient. This opens the route to gene therapy, which involves treating diseases caused by missing or defective genes, and even to repair neurological damage caused by disease, trauma or stroke (Figure 2.22).

The cationic amino groups at the ORMOSIL cage surface readily bind with negatively charged DNA plasmids. The resulting doped material, using plasmids that contained a gene coding for enhanced

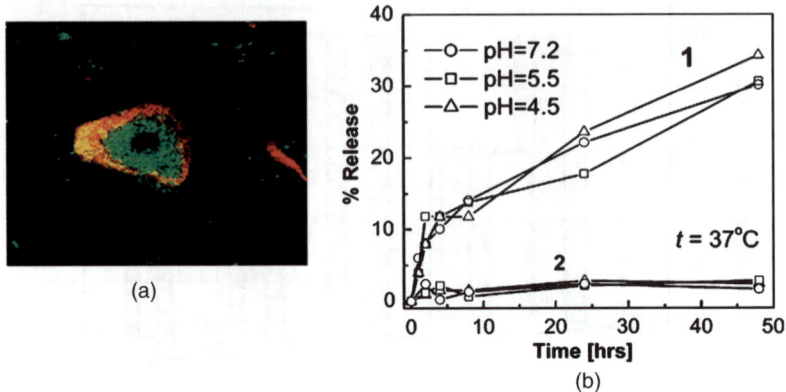

(a)

(b)

Figure 2.22 (a) Nanoparticle-delivered EGFP (green) expressed in a nerve cell (red). (b) The release kinetics of an amphiphilic dye (Rh6G; curves 1) and hydrophobic dye (HPPH; curves 2) from ORMN20 nanoparticles shows that encapsulation of hydrophobic dyes in ORMOSIL nanoparticles can be used for optical tracking of nanoparticle delivery, whereas that of amphiphilic drugs/dyes can be exploited for controlled release. (Reproduced from ref. 16, with permission.)

green fluorescent protein (EGFP), injected into various regions of the mouse brain, was even more effective than some viral vectors at delivering the EGFP gene to nerve cells without any damage to the mice nerve cells (Figure 2.23).

Release of DNA *in vivo* takes place due to the increased acidic conditions inside living cells that result in the destabilization of the ORMOSIL–DNA complex. SiO_2-based nanoparticles, in fact, do *not* release encapsulated biomolecules because of the strong hydrogen bonding between the biomolecule's polar centres and the silanols at the cage surface (as ORMOSIL-entrapped hydrophobic molecules are *not* leached in aqueous systems due to strong hydrophobic interactions).[17]

Hence, by using an amino-functionalized ORMOSIL of intermediate HLB to entrap amphiphilic molecules such as DNA, the plasmids bound to the amino-functionalized nanoparticles are completely protected against enzymatic degradation due to hindered access of the enzymes to the DNA that is immobilized at the cage surface, while a controlled release behaviour, with an initial burst ($\sim 10\%$ release in the first 3 h), followed by a slow release ($\sim 30\%$ release in 48 h) is observed. These experiments clearly demonstrate that ORMOSIL can be used not only for tracking the fate of delivered silica nanoparticles in the brain via delivery of reporter genes such as EGFP, but also for monitoring or even controlling cell biology *in situ* via efficient *in vivo* delivery of clinically relevant genes such as nucleus-specific FGFR-1.[18]

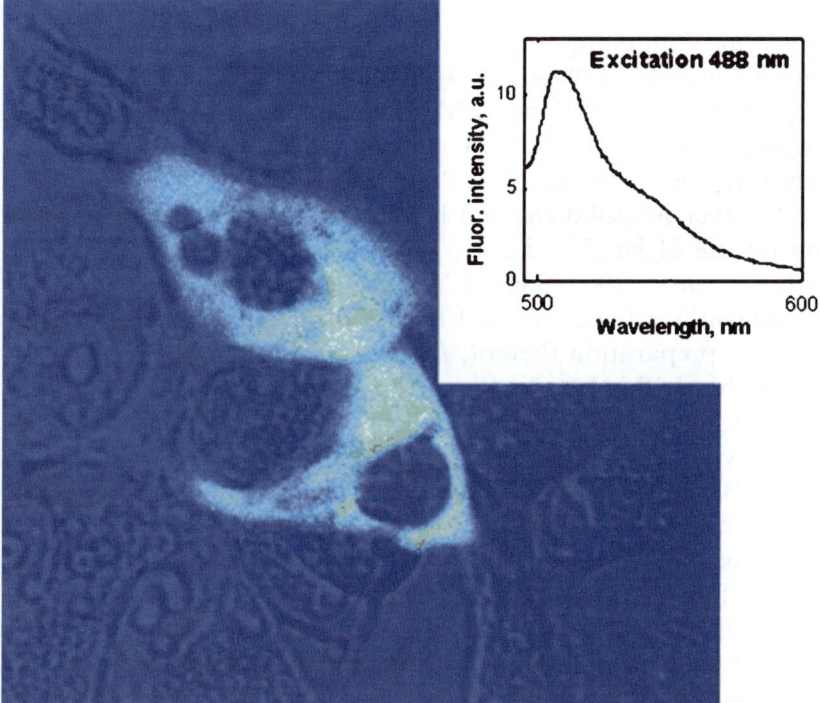

Figure 2.23 Combined transmission and fluorescence (the inset shows the spectrum of EGFP taken from cell cytoplasm) spectra show that cells are effectively transfected with pEGFP when delivered with ORMOSIL nanoparticles. (Reproduced from Ref. 17, with permission.)

This work and other similar studies also make it clear that there is little correlation between *in vitro* and *in vivo* gene delivery using most transfection systems. Traditional ways of testing new DNA delivery agents, by using cultured model cell lines with reporter genes, may need to be re-evaluated and changed. Instead, specific cells that best represent tissue and intended genes should be assessed at a very early stage. A company, Solidus Biosciences,[19] was started up in the USA in 2004 that is now commercializing the technology.

References

1. Downy – Radiance. See: www.discoveringradiance.com.
2. M. McCoy, Encapsulating a new business, *Chem. Eng. News*, 2008, **86**, 26.
3. Several patent applications have been granted to SGT covering its sol–gel microencapsulation process and its implementation for

BPO, sunscreens and other applications. See, for instance, Compositions containing oils having a specific gravity higher than the specific gravity of water, WO/2003/039510, 2003.

4. M. Ahola, E. S. Säilynoja, M. H. Raitavuo, M. H. Vaahtio, J. I. Salonen and A. U. O. Yli-Urpo, In vitro release of heparin from silica xerogels, *Biomaterials*, 2001, **22**, 2163.

5. R. Viitalaa, M. Jokinena and J. B. Rosenholma, Mechanistic studies on release of large and small molecules from biodegradable SiO_2, *Int. J. Pharm.*, 2007, **336**, 382.

6. Compositions for controlled release of a biologically active agent, and the preparation thereof, *US Pat.*, 7112339.

7. C. Barbé, J. Bartlett, L. Kong, K. Finnie, H. Q. Lin, M. Larkin, S. Calleja, A. Bush and G. Calleja, Silica particles: a novel drug-delivery system, *Adv. Mater.*, 2004, **16**, 1959.

8. C. J. Barbé, L. Kong Kim, S. Finnie, S. Calleja, J. V. Hanna, E. Drabarek, D. T. Cassidy and M. G. Blackford, Sol-gel matrices for controlled release: from macro to nano using emulsion polymerisation, *J. Sol-Gel Sci. Technol.*, 2008, **46**, 393.

9. S. Radin, P. Ducheyne, T. Kamplain and B. H. Tan, Silica sol-gel for the controlled release of antibiotics: I. Synthesis, characterization, and in vitro release, *J. Biomed. Mater. Res.*, 2001, **57**, 313.

10. S. Radin, T. L. Chen and P. Ducheyne, Emulsified sol-gel microspheres for controlled drug delivery, *Key Eng. Mater.*, 2007, **330–332**, 1025.

11. Z. Wu, Y. Jiang, T. Kim and K. Lee, Effects of surface coating on the controlled release of vitamin B1 from mesoporous silica tablets, *J. Controlled Release*, 2007, **119**, 215.

12. P. Ducheyne and S. Radin, Xerogel films for the controlled release of pharmaceutically active molecules, WO/2006/115805, 2006.

13. S. Radin and P. Ducheyne, Controlled release of vancomycin from thin sol–gel films on titanium alloy fracture plate material, *Biomaterials*, 2007, **28**, 1721.

14. B. C. Dave, K. Deshpande, M. S. Gebert and J. C. McAuliffe, Osmoresponsive glasses: osmotically triggered volume changes of organosilica sol-gels as a means for controlled release of biomolecules, *Adv. Mater.*, 2006, **18**, 2009.

15. K. Deshpande, B. C. Dave and M. S. Gebert, Controlled dissolution of organosilica sol-gels as a means for water-regulated release/delivery of actives in fabric care applications, *Chem. Mater.*, 2006, **18**, 4055.

16. D. J. Bharali, I. Klejbor, E. K. Stachowiak, P. Dutta, I. Roy, N. Kaur, E. J. Bergey, P. N. Prasad and M. K. Stachowiak,

Organically modified silica nanoparticles: a nonviral vector for in vivo gene delivery and expression in the brain, *Proc. Natl. Acad. Sci. USA*, 2005, **102**, 11539.

17. I. Roy, T. Y. Ohulchanskyy, D. J. Bharali, H. E. Pudavar, R. A. Mistretta, N. Kaur and P. N. Prasad, Optical tracking of organically modified silica nanoparticles as DNA carriers: a nonviral, nanomedicine approach for gene delivery, *Proc. Natl. Acad. Sci. USA*, 2005, **102**, 279.

18. D. Luo and W. M. Saltzman, Nonviral gene delivery: thinking of silica, *Gene Ther.*, 2006, **13**, 585.

19. See: www.solidusbiosciences.com.

CHAPTER 3
Purification and Synthesis

3.1 Size Control, Shape and Purity for Enhanced Separation

New silica gels obtained by sol–gel polycondensation of tetra-ethylorthosilicate (TEOS) or related silanes offer largely superior performance in liquid chromatography (LC) separation of organic compounds, a task for which several thousands tons of silica are employed worldwide by industry. LC devices now rank third behind analytical balances and pH meters in number of installed analytical instruments.[1]

LC systems are used in a wide variety of applications besides the pharmaceutical industry (the heaviest user of high-performance liquid chromatography (HPLC) technology) including environmental studies to measure pesticides and other contaminants in water; food safety testing to quantify drug residues and other impurities; and in biomedical applications to study amino acids, peptides and other bioanalytes. The silane is polymerized to form high-purity, monodisperse silica spheres (Figure 3.1) free from metals that would otherwise interfere with chromatographic separations.

The particles are mechanically strong, have high surface area and feature pore sizes that can be easily tailored. In addition, manufacturers can customize the properties of the material for a particular application by modifying the surfaces with octyl, octadecyl, phenyl and other types of functional groups that elicit separation-enhancing interactions between the components of the solution being analysed and the particles in the column.

Silica-Based Materials for Advanced Chemical Applications
By Mario Pagliaro
© Mario Pagliaro 2009
Published by the Royal Society of Chemistry, www.rsc.org

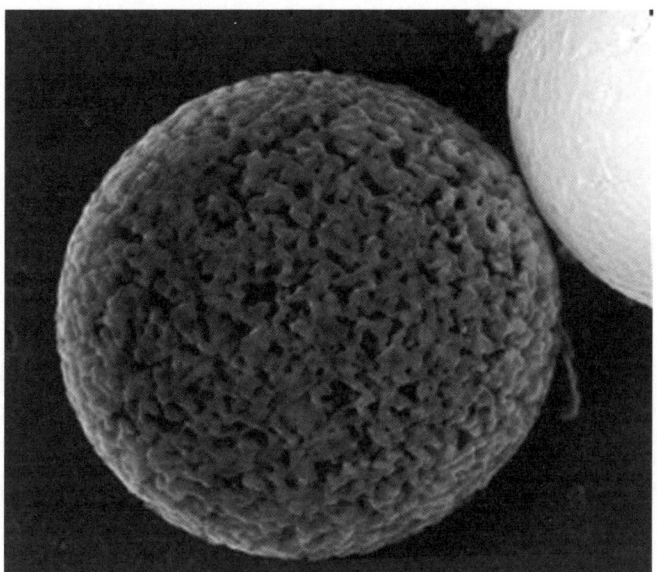

Figure 3.1 By using synthesis methods to control properties such as chemical and structural stability and porosity (dark sphere is porous; white sphere is not), researchers can custom-make separation media for LC applications. These micrometre-sized spheres were made at Waters Corp.

Beyond sol–gel synthesized monolithic silica columns widely employed in the field of high-throughput analysis of drugs and metabolites, as well as in bioanalytical separation,[2] the versatility of the sol–gel process enables the production of new hybrid materials such as ethylene-bridged hybrid materials (Figure 3.2). Using these materials, one can overcome pure silica's tendency to undergo hydrolysis in alkaline environments, rendering the material stable up to a pH of 12, which is an ideal condition for analysing some pharmaceutical agents.

A narrow particle size distribution is crucially important. For example, a high-purity silica gel (Figure 3.3) with a particle size of 40–63 μm compares favourably with the overall flash chromatography industry average of 35–75 μm. This ensures more uniform flash column or cartridge packing, as well as better resolution and separation. Figure 3.4 shows the distribution curves of a commercial gel compared to three other products sold as 40–63 μm gels.

The two key points of Figure 3.4 are the height of the volume differential ("diff.") and percentage of particles below 40 μm (one gel has a mean of 90% of the particles in the nominal range in comparison with 80% for the others). The blue curve has a much higher

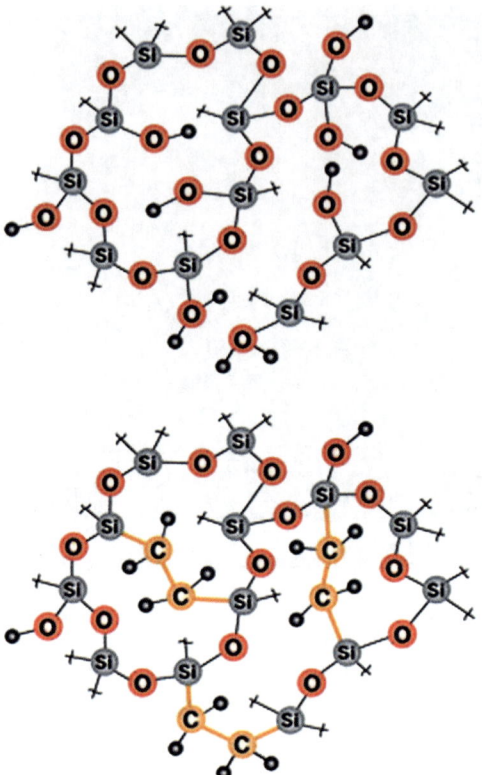

Figure 3.2   Modifying silica particles to include ethylene bridges gives the particles exceptional chemical stability. Si, grey; O, red; H, black; C, orange. The images highlight the difference between silica (top) and ethylene-bridged silica (bottom) particles.

percentage of particles between 40 and 63 μm and a very low level of small particles below 40 μm (or "fines"). Fines increase backpressure that can result in clogging which is particularly dangerous when using glass columns. The fines can also pass through filters and contaminate final products. No fines give a more regular, stable and reproducible chromatography bed that gives a faster and more even flow rate for better separation.

When the particle size range is broad the column packing is uneven. Some parts are composed of only large particles where the solvent will flow fast and meet little resistance, and there are sections composed of small particles where the solvent flows slowly and meets great resistance. As a result, the solvent will take the path of least resistance through the column and flow around the pockets of small particles instead of

Figure 3.3 The particle size of a standard 40–63 μm flash chromatography product (SiliaFlash® F60). (Photo courtesy of SiliCycle.)

straight through the column. This uneven flow greatly affects the separation because the compounds to be separated will have different retention times depending on their flow path through the column. As they exit the column, the compounds will give broad and poorly separated peaks (Figure 3.5).[3]

A narrower particle size distribution will give a more homogenous packing that will help in collecting more concentrated fractions and in reducing solvent consumption, which will ultimately be more cost-efficient. Irregular silica, dependent on its method of manufacturing, normally contains trace quantities of a variety of metals, which in turn can affect the separation. The sol–gel technology based on TEOS polymerization generates silica gels with the lowest trace metal content, and thus with only traces of sodium, iron and lead that otherwise cause peak tailing. Such a high-quality sorbent can thus also be used as a starting material for functionalized silica gels.

Finally, for preparative separation and for certain analytical process separations, materials with spherical particles of monodisperse size (Figure 3.6) are the most useful, despite their high cost. Indeed, the high specific surface area enables a high loading capacity with a uniform and reproducible coverage. Here the sol–gel manufacturing process from an organic form of silicon alkoxide affords spherical silica gels with the desired characteristics. This is achieved through careful monitoring and control of the factors that induce precipitation.

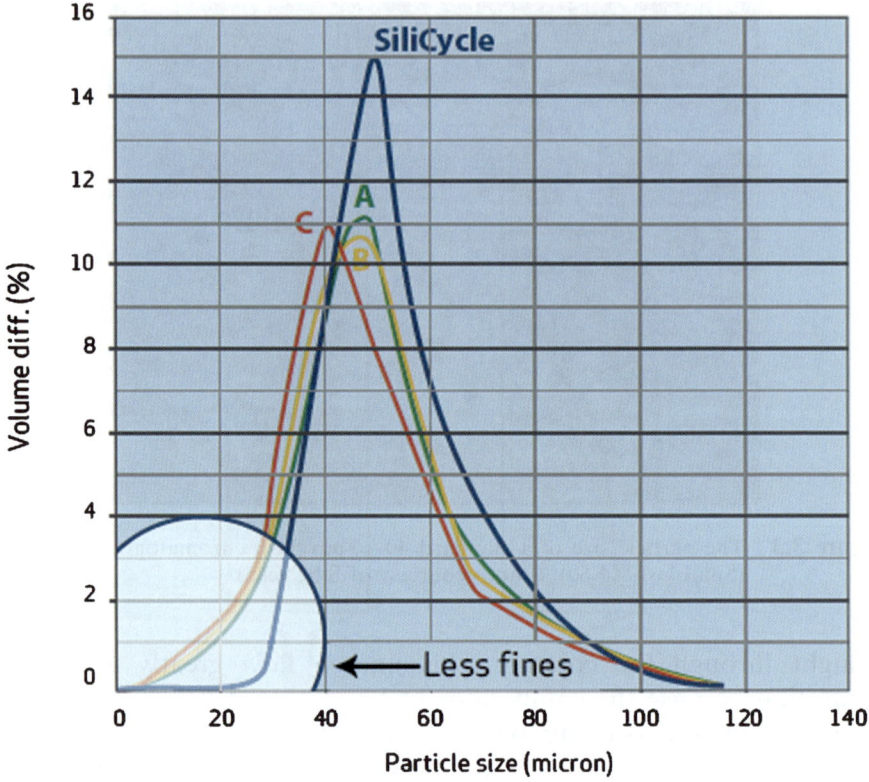

Figure 3.4 Comparison of particle size distribution of four different silica chromatography gels. (Photo courtesy SiliCycle.)

3.2 Scavenging with Functionalized Silicas

The use of solid-supported reagents and scavengers is an excellent way
to expedite organic synthesis by simplifying the purification process.
Scavengers are functionalized silicas designed to react and bind excess
reagents and by-products. The process relies on chemically driven
reactions where the excess reagents and reaction by-products react with
and bind to the scavenger. The solution which now only contains product can be separated from the resin-bound impurities by filtration
(Figure 3.7).

In the case of batch processing the scavenger is added to the reaction
mixture after completion. For flow-through processing the scavenger is
packed into a column (solid-phase extraction, HPLC, *etc*.) and the
reaction mixture is passed through the column until scavenging is
complete.

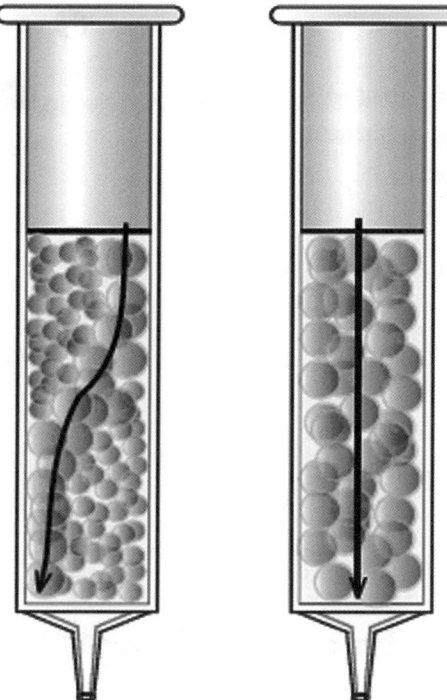

Figure 3.5 Effect of fines (left) on elution compared with a silica gel that has no fines (right).

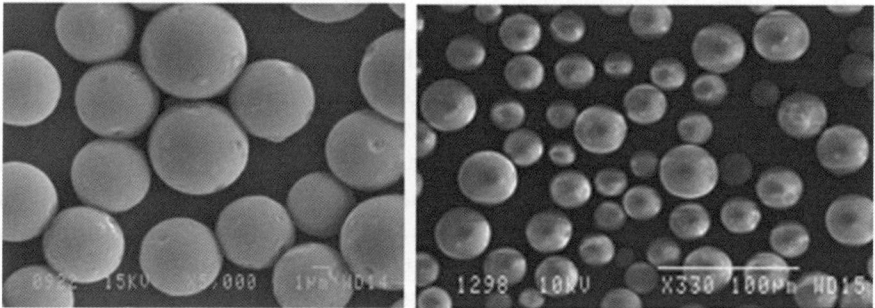

Figure 3.6 Silica spheres with monodisperse particle size are expensive analytical gels of top performance. Silica spheres with determined particle size ranges are ideally suited for preparative chromatography. (Photo courtesy of SiliCycle.)

Standard purification procedures including chromatography, liquid–liquid extraction and crystallization can be time consuming and difficult to scale up. In many cases supported reagents have distinct advantages over their solution-phase counterparts, including increased selectivity,

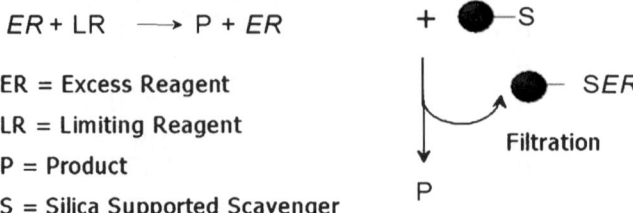

Figure 3.7 The process of using scavengers.

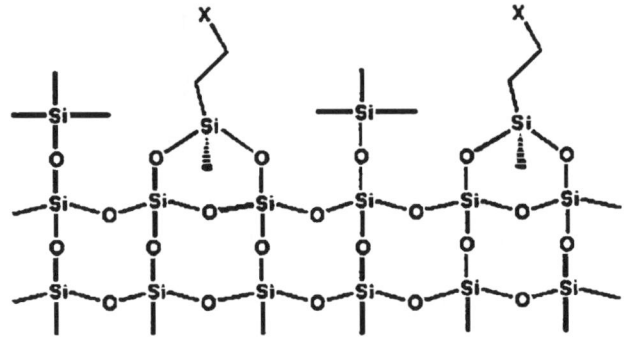

Figure 3.8 The structure of functionalized end-capped silica gel.

the possibility of using incompatible reagents in one "pot" as they cannot react with each other and the immobilization of toxic regents. Reactive functional groups are thus grafted onto silica and then the residual silanol groups are end-capped to produce an inert non-swelling support (Figure 3.8) that usually outperforms polymers.

Functionalized silica gels offer several advantages over traditional functionalized polymers, the major advantage being the inert inorganic backbone that eliminates issues associated with swelling and solvent compatibility. These gels can thus be added directly to a reaction mixture or used in a column to selectively remove metals, making the process scalable all the way from the R&D laboratory up to the process scale, without protocol modifications.

The advantages, in general, include the following.

- *Fast kinetics.* Since the silica is surface functionalized, the rate of reaction is not controlled by the diffusion into and out of the polymer. Most of our scavengers work in under an hour.
- *Solvent independent.* Silica neither shrinks nor swells in any solvent and because it is end-capped it will not dissolve in any solvent.

- *Controlled loading.* Consistent, accurate loading from lot to lot.
- *Mechanically stable.* Works well with overhead stirring and can be used with a bottom coupled spin bar for up to four hours.
- *Thermally stable.* Silica can withstand temperatures of over 200 °C and is suitable for use in microwave synthesizers.
- *Easy to use.* Silica is easy to weigh and handle since it does not carry a static charge and is always free flowing, and thus amenable to automation. High density makes it suitable for small-volume work.
- *Scalable.* Works on the microgram to the many tons scale.
- *Flexible formats.* Because they do not swell, silica gels can be packed into a variety of flow-through formats such as HPLC columns, flash cartridges and well plates.

For example, silica-bound carbonate (or Si–CO$_3$; Figure 3.9) is the silica-bound equivalent of tetramethylammonium carbonate. It can be used as a general base to quench a reaction, to free base amines in their ammonium salt form and to scavenge acids and acidic phenols, including HOBt, which is widely used in amide coupling reactions. It can be tested side by side with MP-Carbonate for the scavenging of HOBt after the following amide coupling reaction:

1) Benzyl amine, HOBt, PS-Carbodiimide, DMA, μW 5 min filter

2) Si-Carbonate, MeOH, 5 min, filter

98% Yield
99% purity
< 15 min

(3.1)

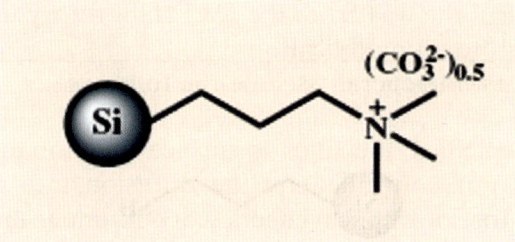

Figure 3.9 Silica-bound carbonate.

The silica-based scavenger has much faster kinetics which allows it to scavenge HOBt in only five minutes (Table 3.1).

Compared to scavenging using the bulk product, scavenging of HOBt over a column packed with silica-functionalized carbonate (SiliaPrep Carbonate) enables optimal sequestration with only 5 ppm of residual peroxide from a 5000 ppm HOBt solution (Table 3.2).[4]

Palladium needs to be scavenged in a number of cases as it is used as a catalyst in widely employed Heck and Suzuki cross-coupling reactions. Compatible with all solvents, organic and aqueous, silica-entrapped thiol (or Si-Thiol; Figure 3.10) is an excellent scavenger (Table 3.3; Figure 3.11). Its mesoporous form, with surface area up to $1200 \, \mathrm{m^2 \, g^{-1}}$, is an excellent adsorbent for environmental remediation, being extremely effective in the removal of mercury (from 0.24 to $1.26 \, \mathrm{mmol \, g^{-1}}$, depending on the degree of functionalization) and silver ions (maximum loading of $0.89 \, \mathrm{mmol \, g^{-1}}$) from aqueous solutions.[5]

Table 3.1 Solid-phase extraction using polystyrene and silica-based supports.

Solid-phase extraction media	Time (min)	HOBt removed[a] (%)
MP-Carbonate	120	77
MP-Carbonate	240	100
Si-Carbonate	5	100

[a]HOBt scavenging determined by LC-MS and verified by ^{1}H NMR.

Table 3.2 Scavenging HOBt with SiliaBond® Carbonate (bulk) and SiliaPrep™ Carbonate.

Initial HOBt concentration (ppm)	Si–CO₃ (equiv.)	Reaction time	Final HOBt concentration[a] (ppm)	Reactivity (%)
5000	3	5 min	32	99.4
5000	3	1 h	32	99.4
5000	3	SiliaPrep™	<5	100.0
5000	4	5 min	22	99.6
5000	4	10 min	22	99.6
5000	4	1 h	21	99.6
5000	4	SiliaPrep™	<5	100.0

[a]HOBt concentration determined by GC-MS; solvent: dimethylformamide.

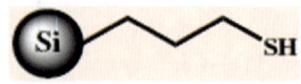

Figure 3.10 Silica-entrapped thiol.

Table 3.3 Scavenging palladium complexes with SiliaBond® Thiol.

Reaction time	Pd(AcO)$_2$	Pd$_2$(C$_3$H$_5$)$_2$Cl$_2$	Pd(PPh$_3$)$_4$	Pd$_2$(dba)$_3$
5 min	0.9	–	360	545
60 min	0.07	0.04	320	20
18 h	0.05	–	150	–

Starting concentration: 1000 ppm. Scavenging with 4 equiv. of SiliaBond® Thiol in tetrahydrofuran at room temperature.

Figure 3.11 The orange colour of Pd(AcO)$_2$ disappears from a solution five minutes after adding silica-supported thiol. (Image courtesy of SiliCycle.)

The nature of the palladium complex or the ligand coordinated to the metal is very important for the reaction itself and for the removal of the metal since the scavenger needs to have a stronger affinity for the palladium than for the ligand to effectively remove it.[6] Strong ligands will hinder the scavenging of the metal much more than weak ones; strength is determined by the size of the ligand and the level of bonding energy that it has towards palladium. Another important factor is the oxidation state of palladium. Pd(0) will be more difficult to scavenge because it has tetrahedral orbitals that will react by the S$_n$1 mechanism which is much slower than the S$_n$2 mechanism that the square planar Pd(II) moiety will undergo. The four functionalized silicas shown in Figure 3.12 were used to scavenge complexes Pd(AcO)$_2$, Pd$_2$(C$_3$H5)$_2$Cl$_2$, Pd(PPh$_3$)$_4$ and Pd$_2$(dba)$_3$. The results in Table 3.4 show that, as predicted, the easiest palladium complexes to remove are the acetate containing Pd(II) and the other ionic complex Pd$_2$(C$_3$H$_5$)$_2$Cl$_2$.

The two Pd(0) complexes, Pd(PPh$_3$)$_4$ and Pd$_2$(dba)$_3$, were not scavenged as effectively overall, but eventually the Si-TAA and Si-Thiourea

Si-Thiol

Si-Thiourea

Si-Triamine

Si-Triaminetetraacetic
Acid
Si-TAA

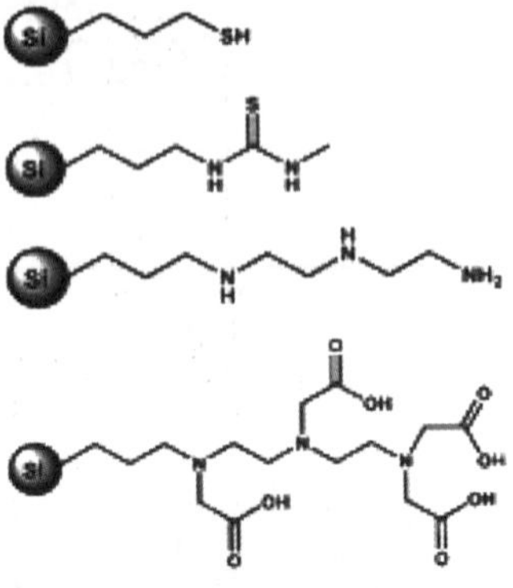

Figure 3.12 Functionalized silicas.

Table 3.4 Scavenging of palladium complexes at room temperature with 4 equiv. in tetrahydrofuran for three reaction times.

Scavenger	Reaction time	$Pd(AcO)_2$	$Pd_2(C_3H5)_2Cl_2$	$Pd(PPh_3)_4$	$Pd_2(dba)_3$
Si-Thiol	5 min	0.90	–	360	545
	60 min	0.07	0.04	320	20
	18 h	0.05	–	150	100
Si-Thiourea	5 min	1.4	–	320	475
	60 min	0.8	1.3	95	50
	18 h	0.6	–	10	90
Si-TAA[a]	5 min	40	–	390	480
	60 min	9.8	0.25	150	50
	18 h	0.06	–	1.4	190
Si-Triamine	5 min	20	–	540	525
	60 min	1.4	1.3	370	83
	18 h	0.3	–	220	280

[a]TAA, triaminetetraacetic acid.

were effective over time. It is interesting to note that with $Pd_2(dba)_3$ the lowest level of palladium was achieved after one hour, due to the reaction slowly reaching equilibrium. It is therefore important to follow the kinetics of any scavenger evaluation to optimize the process. Finally, as with most reactions, the number of equivalents is also very important, as can be seen in Figure 3.13 which plots the palladium level over time when using 1, 2 and 4 equiv. of Si-Thiol to remove $Pd(AcO)_2$. After an hour there is a large increase in the amount of palladium removed with two equivalents *versus* one and slight increase with four *versus* two. Over time the effect decreases, and a balance must be struck between the time savings and efficiency and the increased cost of adding additional scavenger.

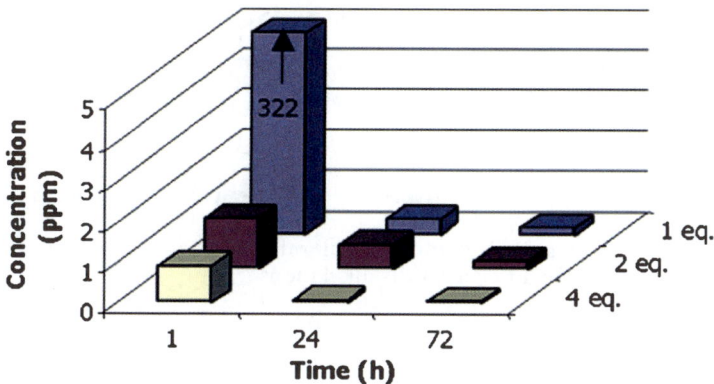

Figure 3.13 Scavenging of palladium using three different quantities of Si-Thiol scavenger. (Complex: $Pd(AcO)_2$; solvent: tetrahydrofuran; scavenger: Si-Thiol; initial concentration: 1000 ppm; room temperature.) (Reproduced from ref. 6, with permission.)

3.3 Silica-Entrapped Reactants

A supported reagent, in a manner similar to a scavenger, consists of reactive functional groups grafted onto insoluble silica. Unlike a scavenger, which is added once a reaction is complete, a supported reagent is added at the beginning of the reaction and replaces the solution-phase reagent, allowing use of the entrapped reagent in great excess to drive the reaction to completion. The spent reagent is easily removed by filtration (Figure 3.14).

Bound reagents are an excellent alternative in cases where the reagent is used in excess and can be difficult to remove. Purification is now a simple process of filtration and evaporation. For example, silica-supported aluminium chloride is a Lewis acid and an effective catalyst for Friedel–Crafts alkylations (Figure 3.15).

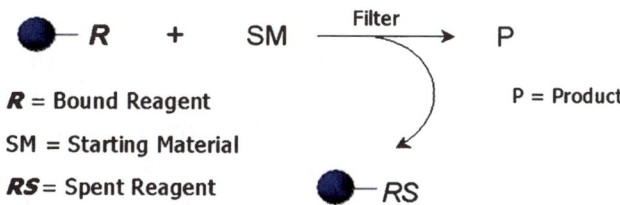

Figure 3.14 A supported reagent, in a manner similar to a scavenger, consists of a reactive functional group grafted onto insoluble silica. The spent reagent is easily removed by filtration.

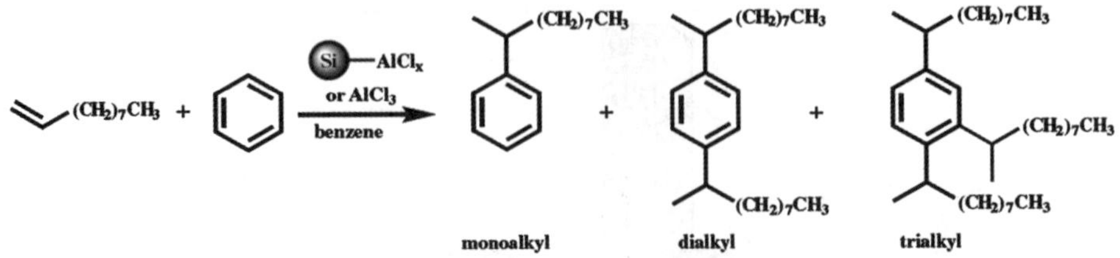

Figure 3.15 Silica-supported aluminium chloride is a Lewis acid and an effective catalyst for Friedel–Crafts alkylations.

Table 3.5 Formation of linear alkylbenzene by Friedel–Crafts alkylation.

Alkene	Catalyst	Alkene conversion (%)	Selectivity towards alkylbenzene (%)		
			Mono	Di	Tri
1-Hexene	$AlCl_3$	100	58.6	31.1	10.3
1-Hexene	$Si-AlCl_x$	100	71.0	28.0	1.0
1-Decene	$AlCl_3$	100	68.5	22.5	9.0
1-Decene	$Si-AlCl_x$	100	80.0	20.0	–

The formation of linear alkylbenzene (LAB) by Friedel–Crafts alkylation is strongly influenced by homogeneous or heterogeneous reaction conditions (Table 3.5).[7] The steric bulk of silica reduces over-alkylation and shelf life. Silica also catalyses the formation of ethers.

Similarly, heterogeneous amide synthesis can be elegantly achieved in cartridges filled with silica functionalized with carbodiimide. The process outlined in Figure 3.16 yields quantitative conversion and 98.8% purity by reacting benzoic acid with phenylethylamine.

Supported reagents can also be used for purification. "Catch and release" is a technique commonly used to purify amines using a silica-supported acid (SCX or SCX-2, manufactured by SiliCycle). In the procedure (Figure 3.17), the silica-supported acid "catches" the amine of interest and then all of the impurities are washed out. Once this is done, the amine is "released" from the supported acid with a solution of 2 M of ammonia in methanol. This technique can also be used to purify carboxylic acids using a silica-supported amine (SiliaBond Amine) and releasing with a 5% acetic acid in methanol solution. SiliCycle's SCX, SCX-2 and SiliaBond Amine have consistently higher loadings and purity, making them ideal for these applications.

SiliCycle manufactures a number of functionalized silica gels to meet the specific requirements of pharmaceutical professionals, and has

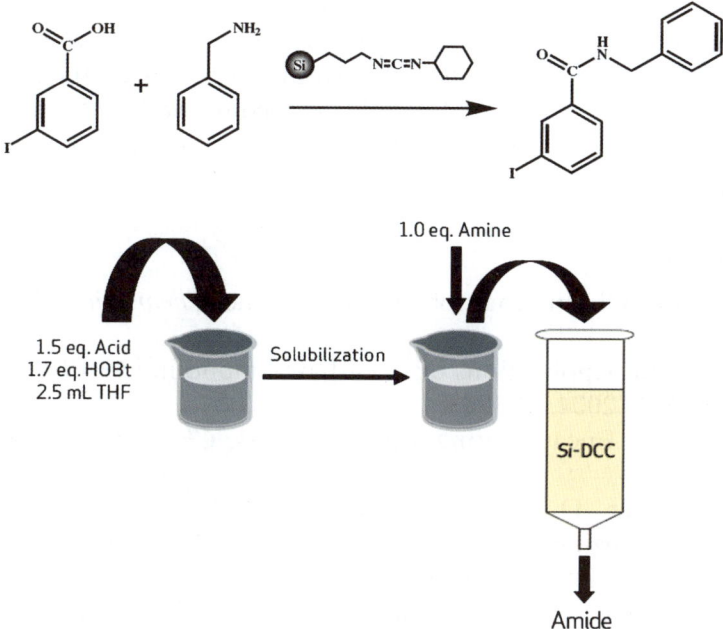

Figure 3.16 Using SiliaPrep™ Carbodiimide for amide synthesis.

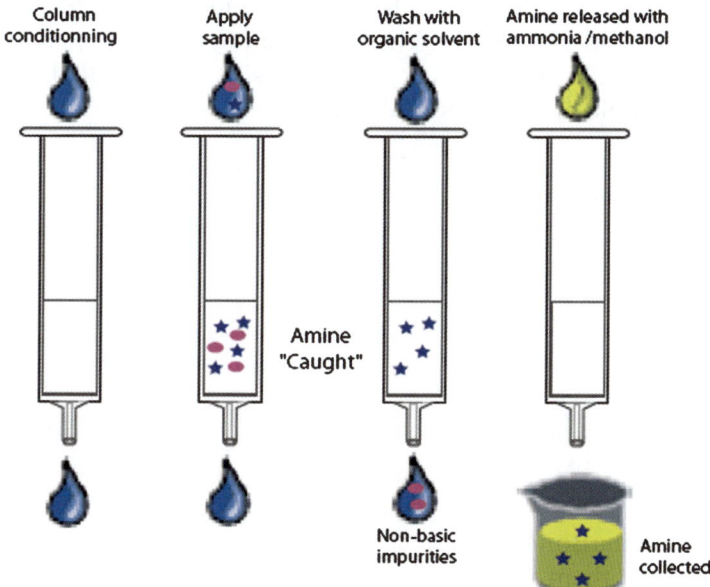

Figure 3.17 Amine "catch and release" purification with a silica-supported acid.

launched a line of mesoporous materials with further advantages in terms of performance selectivity and process speed. Showing the relevance of silica-based synthesis and purification processes, the company's revenues in 2007 increased fourfold, exceeding $8.5 million, and the company is rapidly growing in terms of personnel, production plants and market share.[8]

References

1. M. Jacoby, Chromatography in the extreme, *Chem. Eng. News*, 2008, **86**, 17.
2. K. Cabrera, Applications of silica-based monolithic HPLC columns, *J. Sep. Sci.*, 2004, **27**, 843.
3. M. McCoy, Encapsulating a new business, *Chem. Eng. News*, 2008, **86**, 26.
4. D. R. Sauer, D. Kalvin and K. M. Phelan, Microwave-assisted synthesis utilizing supported reagents: a rapid and efficient acylation procedure, *Org. Lett.*, 2003, **24**, 4721.
5. R. I. Nooney, M. Kalyanaraman, G. Kennedy and E. J. Maginn, Heavy metal remediation using functionalized mesoporous silicas with controlled macrostructure, *Langmuir*, 2001, **17**, 528.
6. This study can be found at: www.silicycle.com/html/english/research/pd.php.
7. X. C. Hu, M. L. Foo, G. K. Chuah and S. Jaenicke, Pore size engineering on MCM-41: selectivity tuning of heterogenized AlCl$_3$ for the synthesis of linear alkyl benzenes, *J. Catal.*, 2000, **195**, 412.
8. Data available at: www.hoovers.com.

CHAPTER 4
Coatings

4.1 ORMOSIL-Based Coatings

Given the ease of forming films by traditional wet deposition techniques such as spin-, dip-, flow- and spray-coating, sol–gel silicas have been used as commercial coatings since the 1950s when zinc-rich inorganic paints (still viable commercial products) formulated from oligomeric condensates of tetraethylorthosilicate (TEOS) and zinc powder were first applied as corrosion-resistant and high-temperature coatings for steel.[1]

Today, hybrid organically modified silicate (ORMOSIL) coatings are of special interest since their properties, intermediate between those of polymers and glasses, can meet the specific and unique requirements of material properties not afforded by organic polymers and glasses alone. This is particularly the case for doped ORMOSIL, the development of which has opened the route to *multifunctional* coatings that are finding applications in many diverse fields including cultural heritage protection (Figure 4.1). In general, two approaches can be pursued. In the first, more recently explored, the physical and chemical properties of the coating are widely tailored by changing the non-hydrolysable moiety R in the ORMOSIL:

$$RSi(OCH_3)_{3(aq)} + 3H_2O_{(l)} \xrightarrow{H^+/OH^-} RSi(OH)_{3(aq)} + 3CH_3OH_{(aq)} \qquad (4.1)$$

$$nRSi(OH)_{3(aq)} \xrightarrow{H^+/OH^-} (RSiO_{1.5})_n + 1.5nH_2O_{(l)} \qquad (4.2)$$

Silica-Based Materials for Advanced Chemical Applications
By Mario Pagliaro
© Mario Pagliaro 2009
Published by the Royal Society of Chemistry, www.rsc.org

Figure 4.1   Artificial ultraviolet light can cause valuable artworks to fade, but ORMOSIL-based coatings are highly effective in protecting cultural heritage. (Reproduced from RSC.org)

For example, the mechanical properties—hardness and elastic modulus—of ORMOSIL can be tuned by varying the degree of alkylation and thus the fraction of six- and four-member siloxane rings in the organosilica matrix. This enables fine tuneability of parameters of crucial practical importance (Figure 4.2).[2]

In the second, complementary, approach the polycondensation of silica polymer is followed by the formation of an organic network made by cross-linking reaction of monomers covalently bound to silicon compounds (Scheme 4.1) resulting in polymeric materials with outstanding protective abilities, including thermal, mechanical and corrosion resistance.

For example, a sol–gel coating comprising fluoropolymer particles affords scratch-resistant, anti-stick and low-friction coatings commercialized by DuPont,[3] in which the fluoropolymer particles are distributed homogeneously in the coating. These coatings are suitable, for example, for electrical and non-electrical domestic appliances. Given the great versatility of the sol–gel process, companies operating in the coating business in general offer custom material design services to find the best hybrid material for specific needs.

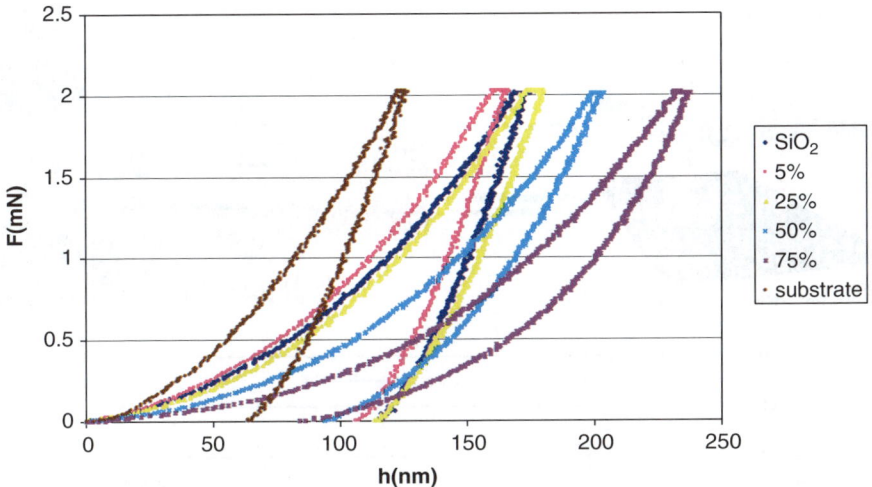

Figure 4.2 Loading curves for ORMOSIL thin films with varying degrees of alkylation of the silica matrix. (Reproduced from ref. 2, with permission.)

1st step: Formation of the inorganic network

organically modified Si alkoxides
Hydrolysis (+H_2O)   →   ≡Si-OH
Polycondensation   →   ≡Si-O-Si≡
(Cocondensation with other metal alkoxides)

2nd step: Formation of the organic network

≡Si-X   +   X-Si≡   →   ≡Si≈ ≈ ≈Si≡
Crosslinking reactions of monomers X covalently bound to Si compounds:
acrylic, vinyl, epoxy, etc.

Curing: thermal, UV/IR, redox initiation

Scheme 4.1 General process of ORMOSIL production.

4.2 Optical Coatings

Porous SiO_2 nanoparticles deposited by a common wet-coating sol–gel technique, for instance, form a single-layer, low-refractive-index and cost-efficient antireflective coating on glass as an alternative to common multilayer compositions (Figure 4.3).[4] The coating thereby obtained exhibits porosity of up to 50% and a refractive index between 1.25 and 1.3, which corresponds to a transmission maximum up to 99% and an

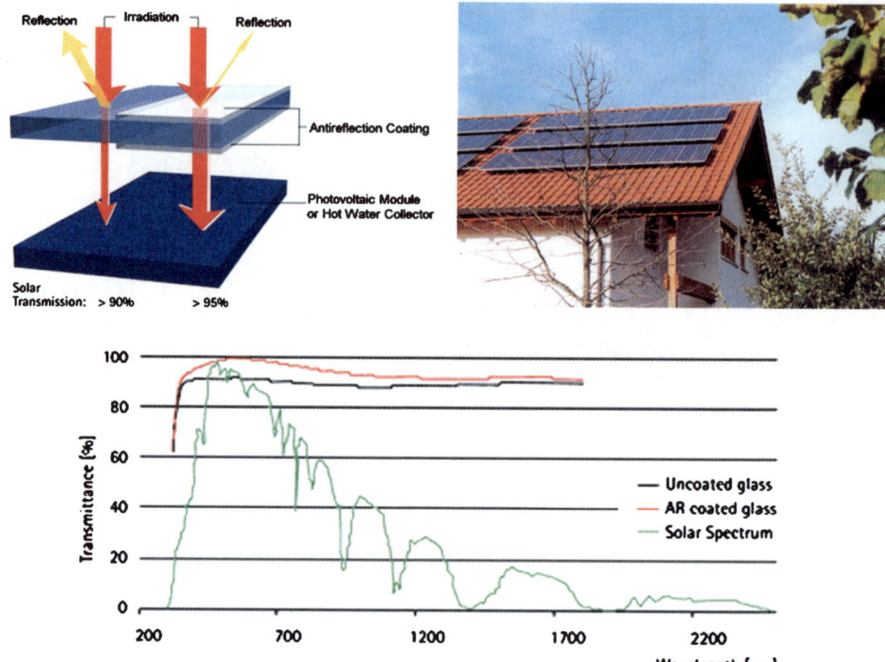

Figure 4.3   Transmission benefit of porous SiO_2 antireflection layer. The solar transmittance of a glass sheet increases from 90 to 95%. (Reproduced from Merck.com)

average solar transmission (weighted over total solar range) $>95\%$. Solar applications (*e.g.* glass cover sheets for photovoltaic modules and hot water collectors) are obvious as well as architectural glazing that requires antireflective glass with certain colour neutrality. When these coatings are used in photovoltaic applications, a watt-peak extra of 3.5% of the original performance can be achieved.

One of the most important drawbacks of classic and new advanced functional materials for applications outdoors, or in environments with high ultraviolet (UV) irradiation, is the light-induced damage that reduces drastically the effective operation lifetime or durability. UV light, either natural or artificial, causes organic compounds to decompose and degrade, because the energy of the photons in UV light is high enough to break chemical bonds.

Organic materials, such as polymers, paints, pigments and dyes, are used in everything from car parts to fine art. Polymers exposed to UV light can lose mechanical strength and integrity, while UV light causes

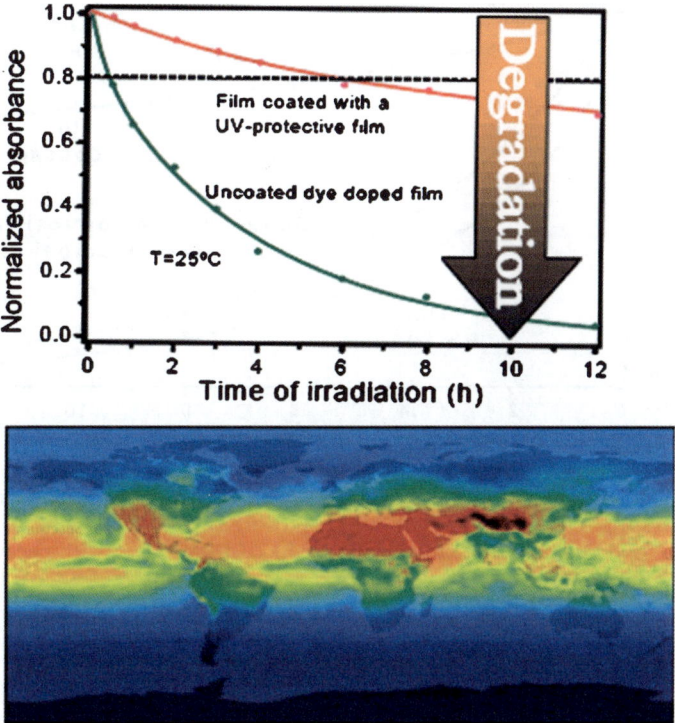

Figure 4.4 Solar UV levels vary across the globe, and depend on time of day and season. The highest levels are shown in dark red. (Reproduced from RSC.org)

the cellulose and lignin in wood to degrade, discolouring the wood and eventually causing fractures and cracking. The dyes in paintings and photographs progressively fade under UV light and paper becomes yellowed and brittle. UV light is the main factor responsible for the degradation of wooden furniture, plastic parts used in the car industry and artwork in museums, which are all exposed to natural or artificial lighting for long periods of time. This makes protecting light-sensitive materials against UV irradiation an important technological demand in almost every industrial field (Figure 4.4).

Sol–gel ORMOSIL materials are excellent for use in applications to protect against UV radiation. ORMOSIL-based coatings made of large amounts of organic UV absorber molecules entrapped in modified silica matrices are capable of reducing drastically UV light reaching a substrate that needs to be protected, and hence reducing the photo-degradation of the substrate upon prolonged exposure to UV sources.[5]

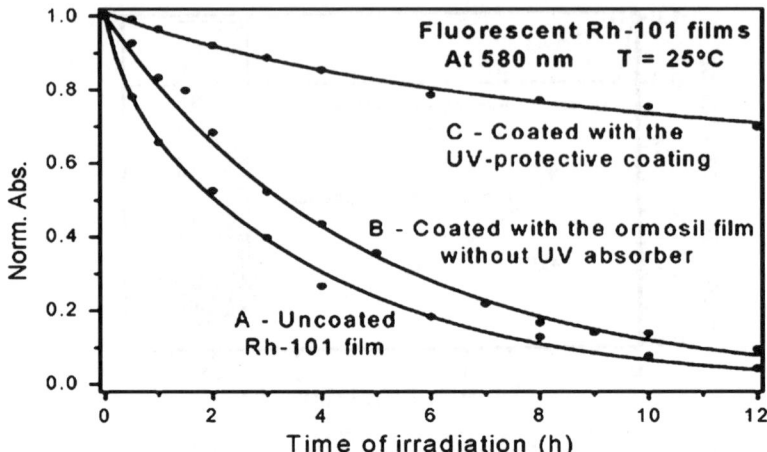

Figure 4.5   Photodegradation of rhodamine films: (A) uncoated; (B) coated with an ORMOSIL coating without UV absorber; (C) coated with a UV-protective ORMOSIL coating.[6] (Reproduced from ref. 6, with permission.)

For example, a thin film of phenylsilica doped with rhodamine dye obtained from a precursor solution consisting of mixtures TEOS and phenyltriethoxysilane (PhTES) and ethanol in a 0.7 : 0.3 : 1 molar ratio reduces the UV light reaching the substrate to less than 7% of the incident light. The degradation of 20% of the dye molecules is 14 times slower for coated samples. Figure 4.5 shows the results of exposing coated and uncoated samples to intense UV radiation, and the intensity of the absorption maximum as a function of the irradiation time. The photodegradation of the dye molecules in the coated samples was much slower compared to the uncoated samples at 25 °C. The UV-absorbing molecules are mainly responsible for the absorption in the UV range, as the degradation of the fluorescent Rh-101 film coated with an ORMOSIL film with the same composition as the protective coating *without* the UV absorber molecules is slightly slower (due to the reflection of light on the surface of the protective coating) than that of the uncoated Rh-101 film. These coatings are just 1 μm thick.

The coatings are also highly stable upon prolonged exposure to UV light and are fully transparent in the visible region of the spectrum. This means they can be used to coat a wide range of materials, without affecting the way they look. The ability to increase the durability of outdoor products that can withstand solar radiation for months or years by a factor of 14 (Figure 4.1) makes these types of protective coatings very attractive for commercial applications.

4.3 Antifouling Coatings

New, potent antifouling coatings are simply based on the hydrophobicity imparted to immersed surfaces by ORMOSIL coatings. In particular, reduced *Ulva* (syn. *Enteromorpha*) zoospore settlement, increased removal of zoospores, increased removal of *Ulva* biomass and fouling release of juvenile barnacles of *Balanus amphitrite* have been achieved with xerogel surfaces of low wettability and low critical surface tension.[7] Hence, hybrid sol–gel derived xerogel films prepared from 50 : 50 *n*-octyltriethoxysilane–tetramethylorthosilane (C8-TEOS–TMOS) with low critical surface tension inhibit settlement of zoospores of the marine fouling alga *Ulva* as well as of juveniles of the tropical barnacle *B. amphitrite*.

Exposed to pressure from a water jet or to turbulent flow generated by a flow channel, such a film gives significantly greater release of eight-day *Ulva* sporeling biomass than other hybrid xerogel surfaces or acid-washed glass. Showing the general validity of the approach to the development of marine antifouling coatings, all other ORMOSIL films tested derived from 45 : 55 (mole ratio) *n*-propyltrimethoxysilane (C3-TMOS)–TMOS, 2 : 98 (mole ratio) bis[3-(trimethoxysilyl)propyl]ethylenediamine (enTMOS)–TEOS and 50 : 50 (mole ratio) 3,3,3-trifluoropropyltrimethoxysilane (TFP-TMOS)–TMOS inhibit settlement of zoospores of marine fouling algae. As shown in Figure 4.6a, the number of spores that settled on all four xerogel films was less in comparison to the glass standard. The settlement density on all four xerogel films was significantly less than the settlement density on the glass control. None of the leachates from the xerogel films were toxic to *Ulva* zoospores. On exposure to a surface stress of 64 kPa water pressure delivered by a water jet (Figure 4.6b), 45–79% of the settled spores were removed by the water jet from the xerogel surfaces as against 33% from glass.

X-Ray photoelectron data indicate that the alkylsilyl residues of the C3-TMOS-, C8-TEOS- and TFP-TMOS-containing xerogels are located on the surface of the xerogel films, which contributes to the film hydrophobicity. The incorporation of the organic functional groups in the hybrid xerogels reduces the available cross-linking in the silicate structure from $Si(OSi)_4$ in the pure TMOS or TEOS to $RSi(OSi)_3$ in the hybrid xerogels leading to a more flexible, less friable surface. Accordingly, the pure TMOS film is highly cracked and poorly adherent, whereas the hybrid xerogel film surfaces are more uniform and uncracked (Figure 4.7).

The xerogel films can be tailored to provide surfaces of different wettability and critical surface tension, and the approach has been patented[8] in the development of a product (Aquafast) that was commercialized in early 2008. The product's usefulness and versatility is

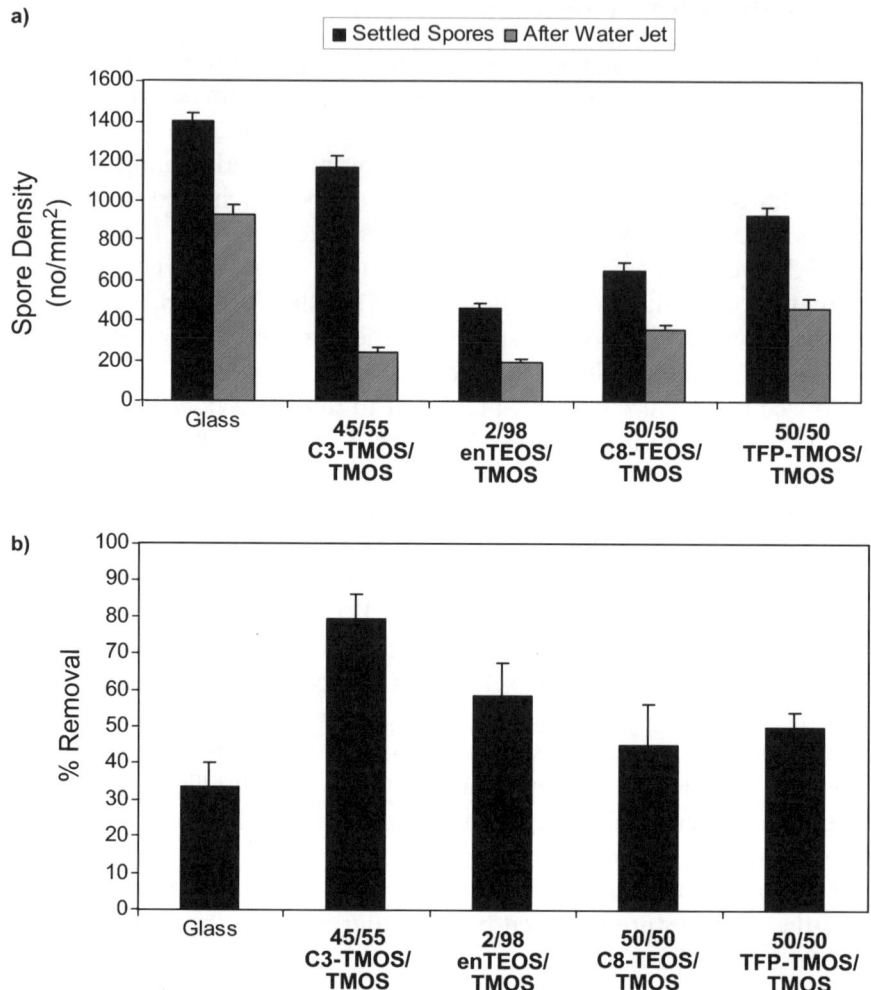

Figure 4.6   (a) The settlement of *Ulva* zoospores on sol–gel coatings and the number of spores remaining after exposure to a water-jet surface pressure of 64 kPa. Each histogram bar is the mean from 90 counts, 30 from each of three replicate slides. Error bars show 95% confidence limits. (b) Percentage removal of zoospores following exposure to the water jet. Each histogram bar represents the mean percentage removal of zoospores from 90 observations of controls (30 from each of three replicate slides) and 90 observations of treatments (30 from each of three replicate slides). Error bars represent 95% confidence limits from arcsine transformed data. (Reproduced from ref. 7, with permission.)

shown, for instance, by its employment to protect the inox steel supporting a video camera 30 m underwater in the sea of the island of Pantelleria (Italy).[9] Painted on with a normal brush (Figure 4.8a), Aquafast successfully protected the device from fouling over the

100 μm

Figure 4.7 Scanning electron microscopy images of TMOS and 50 : 50 C8-TEOS–TMOS xerogel films. (Reproduced from ref. 7, with permission.)

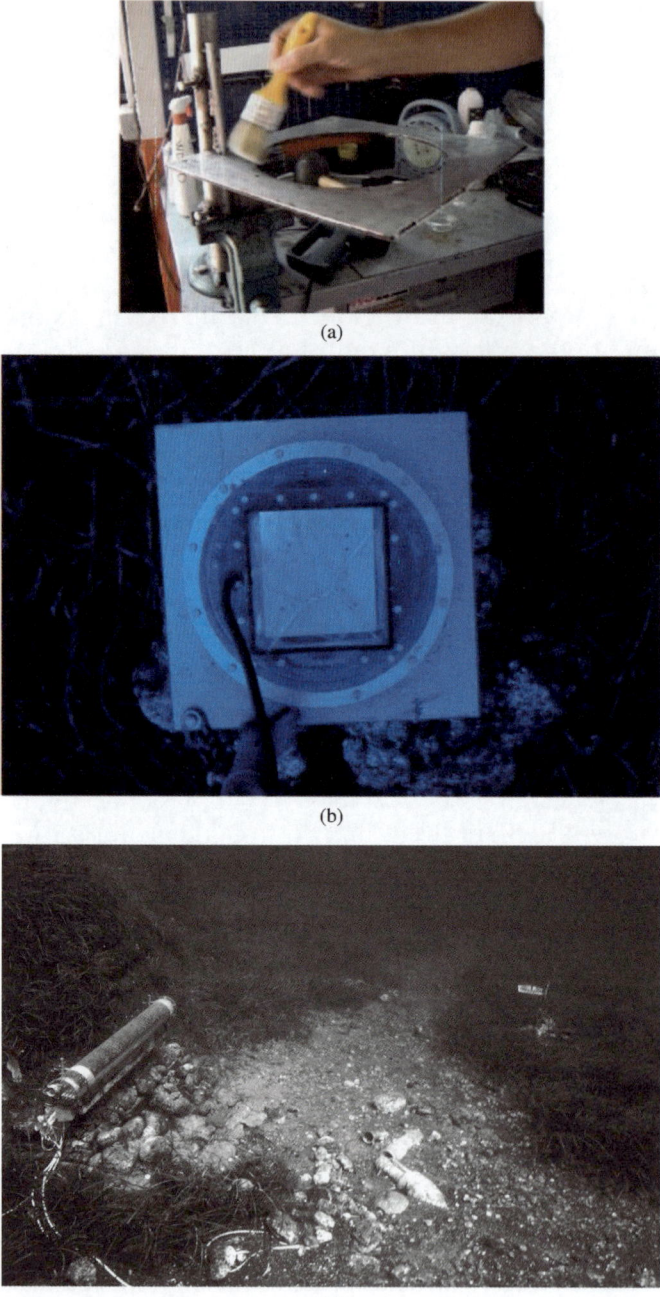

(a)

(b)

(c)

Figure 4.8   (a) Aquafast has been successfully used to protect from fouling the camera devices (b) in the marine environment of Pantelleria Island (Italy) and (c) used to monitor the archaeological site of Cala Gadir. (Photo courtesy of Pietro Selvaggio.)

August–December 2007 period as seen from a picture taken four months after its immersion clearly showing that no vegetation was present (Figure 4.8b).

Aquafast has been applied to the camera used by Sicily's Superintendence of the Sea to monitor the archaeological site of Cala Gadir (Figure 4.8c) and the product is now finding application as an antifouling coating for all the cameras installed by this authority in the sea around Sicily.

The same principle of using low surface energy sol–gel paint is at basis of the BioFlow SAFE paint developed by Safe Marine Nanotechnologies.[10] Again, this commercial paint prevents adhesion of vegetation on ships' hulls without releasing poisonous chemicals into the water. The formulation is mostly made of epoxy–silica hybrids containing small amounts of a silane-functionalized perfluoroether oligomer which reduces considerably the surface energy of the coating due to the migration of fluoroligomer-derived species from the bulk to the surface.[11] Its application is extremely easy (does not differ from that of traditional epoxies), and the paint, which can be supplied transparent or in a selection of colour shades, also crucially reduces a vessel's friction in water.

4.4   Anticorrosion Coatings

Epoxy–silica hybrids are well known for their abrasion resistance and low thermal expansion due to the presence of nanostructured bi-continuous domains. Most recently they have entered the market as anticorrosion coatings in the marine field, for yachts and for large metal vessels, such as oil tankers, and in particular for cargo or ballast tanks and on hulls (Figure 4.9).

Typically, these paints are epoxy–silica hybrids doped with molybdate, which, taking part in the sol–gel condensation reactions, will be part of the inorganic network (Figure 4.10), to be later slowly released as a corrosion inhibitor. Typically, to facilitate the formation of finely dispersed bi-continuous phases the epoxy resin is functionalized with alkoxysilane groups.[12] Then the two components—silica and the epoxy resin—behave synergistically in the hybrid, with the inorganic component enhancing the network density by imposing strong constraints on the molecular relaxations of the organic network. Formed in the early 2000s on the basis of the work developed by a team of Anglo-Italian researchers, SAFE Marine Nanotechnologies today produces a full range of self-healing anticorrosives based on the company's proprietary

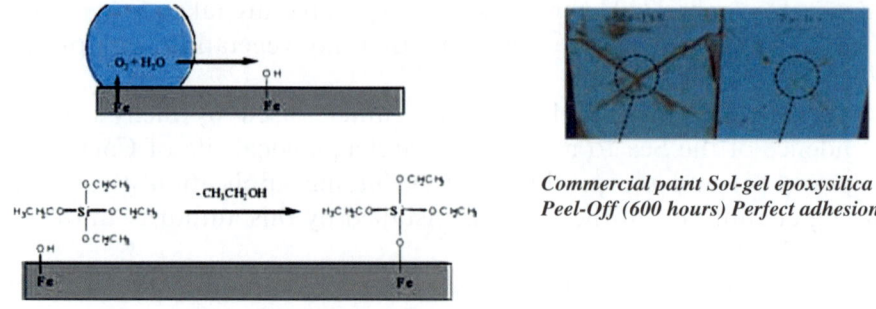

Commercial paint Sol-gel epoxysilica
Peel-Off (600 hours) Perfect adhesion

Figure 4.9   Silicon alkoxides adhere to steel surfaces by chemical binding to surface hydroxyls (left) resulting in strong adhesion of a ballast paint (right). (Reproduced from ref. 12, with permission.)

sol–gel technologies. The result is a group of nanomaterials with unique characteristics: high solids content (up to 98%), solventless, good mechanical and chemical resistance, high adhesion (>25 MPa) and high gloss (>90).

In general, the application of sol–gels as a means of corrosion inhibition is based on two complementary functions: the physical barrier of the polymeric matrix as well as the specific corrosion inhibition of the functional group in the ORMOSIL (the non-hydrolysable moiety R in eqn (4.1) and (4.2)). Magnesium, for example, has high electrical and thermal conductivity, is abundant and easily recycled, and its high strength to weight ratio makes it a valuable asset in the transportation and aviation industries. However, its utilization has not reached full potential because of its high chemical reactivity and tendency to corrode. The very negative standard potential of magnesium ($E^\circ=-2.37$ V *versus* NHE) makes it unstable with respect to water which results in the formation of a protecting oxide layer under ambient conditions. However, the native oxide layer does not prevent its pitting corrosion, facilitated mostly by halides, which results in destructive effects.

Thin sol–gel ORMOSIL–zirconia films successfully inhibit magnesium corrosion (Figure 4.11). The film obtained by combining sol–gel monomers phenyltrimethoxysilane (PTMOS) and zirconium(IV) tetra-1-propoxide (ZrTPO) exhibits superior corrosion inhibition as compared with other films,[13] while the ZrTPO-based film alone does not show significant corrosion inhibition, and the PTMOS-based film provides only moderate protection. The films are prepared by the traditional acid- or base-catalysed hydrolysis and condensation, depositing first the PTMOS film followed by the ZrTPO-based film.

There are hundreds of commercially available trialkoxy-substituted monomers bearing a wide variety of functional groups, which largely

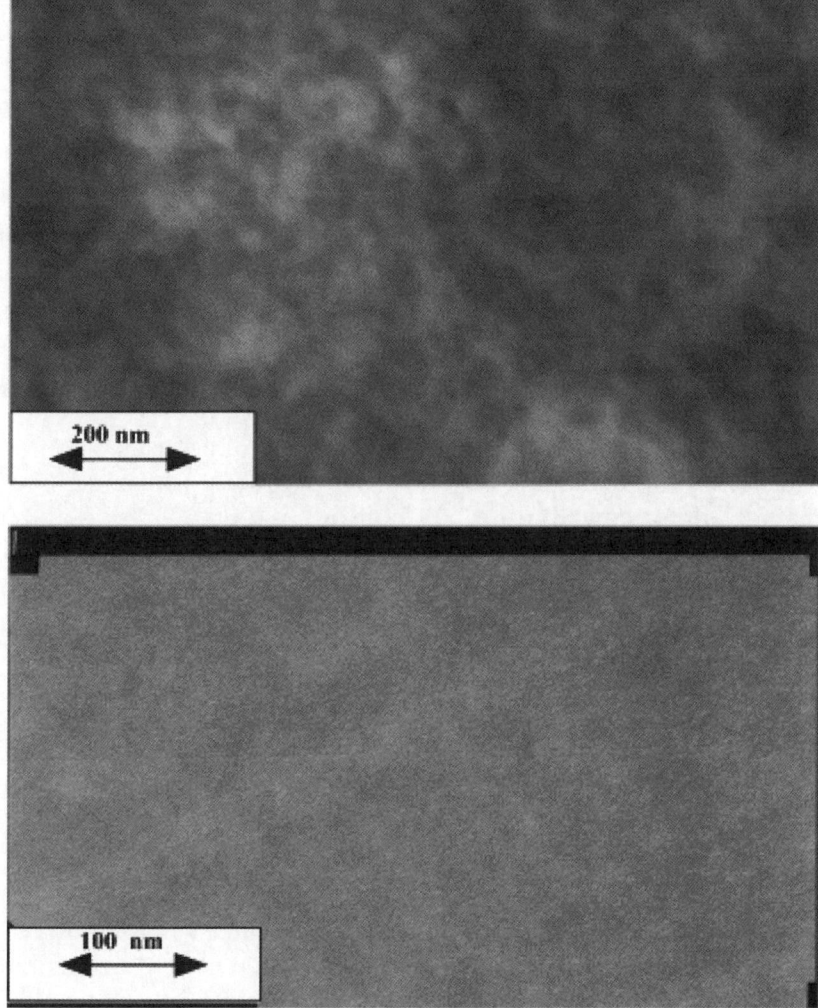

Figure 4.10 Transmission electron micrographs of (top) epoxy–silica hybrid (15% w/w SiO$_2$) and (bottom) the corresponding molybdenum-doped hybrid. (Reproduced from ref. 12, with permission.)

influence the physical and chemical properties of the deposited polymer. For example, sol–gel coatings with phosphonate functionalities also show an improved corrosion protection as compared with pure silica sol–gel coatings due to the strong chemical bonding of phosphonate groups to magnesium substrates.[14] Sol–gels based on 3-methacryloxy-propyltrimethoxysilane and 3-mercaptopropyltrimethoxysilane showed an enhanced corrosion inhibition due to an anodized layer underneath the sol–gel film and its blocking properties.[15]

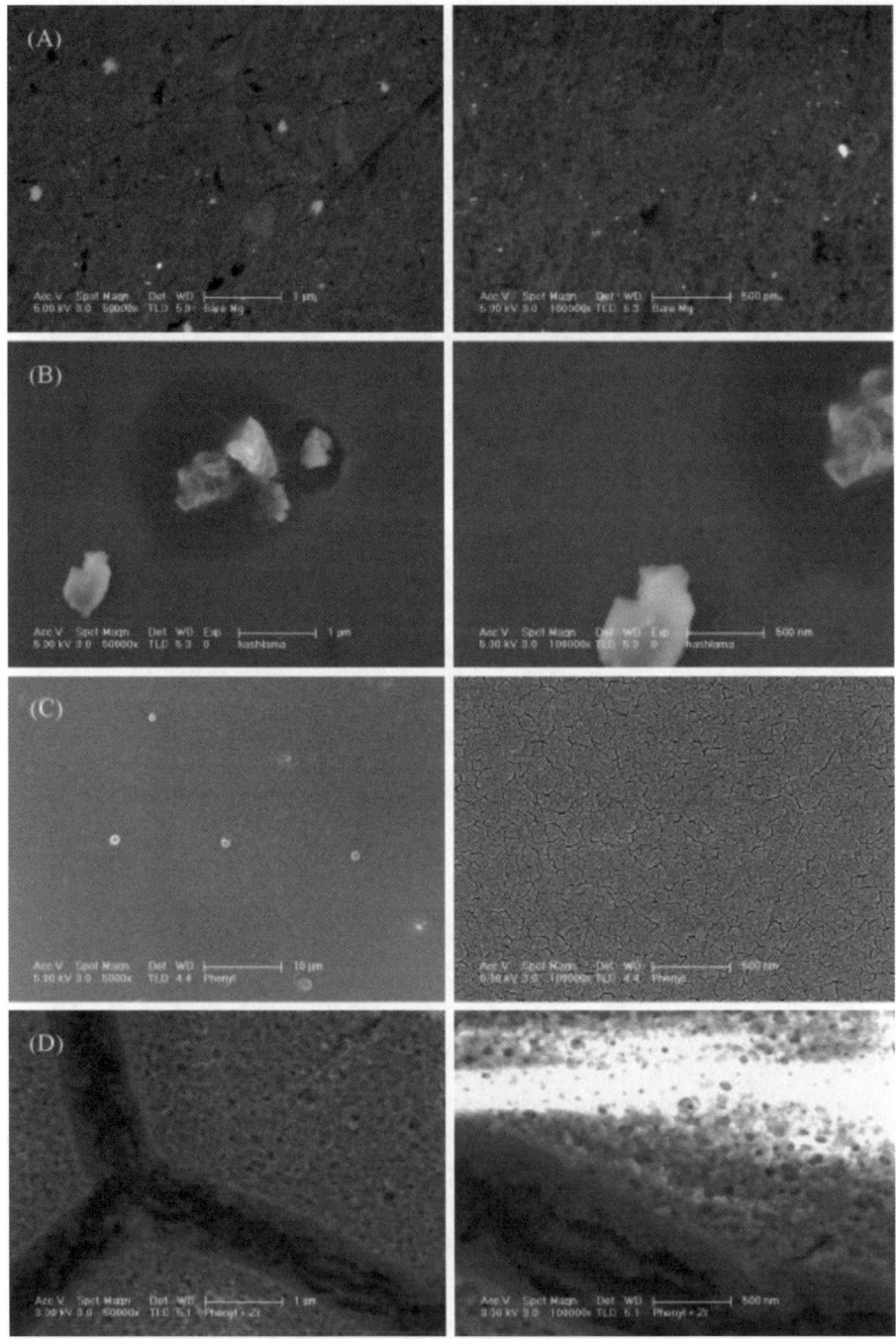

Figure 4.11 Scanning electron microscopy images of bare and coated magnesium chips. Clear morphological differences are seen between the coated and uncoated samples. (Reproduced from ref. 13, with permission.)

An important factor of film quality concerns its adhesion to the substrate, namely to the magnesium chip. Figure 4.12 shows images of different coated chips after a standard adhesion test. The upper images show the magnesium-coated chips after the tape was peeled off, while the lower images show the tapes.

Good adhesion of the layer is necessary, but not sufficient, to inhibit corrosion. Indeed, all three films show good adhesion to the surface (the films were scratched to visualize the removal effect, and to increase the

Figure 4.12 Optical images of coated magnesium chips and their corresponding tapes after an adhesion test: (A) PTMOS film; (B) ZrTPO film; (C) combined film. The images of the tapes are somewhat misleading as the coloured area of the combined film is the largest, which is due to its high content of Rhodamine B. (Reproduced from ref. 13, with permission.)

clarity of the images Rhodamine B was added to the deposition solutions), which cannot account for the difference in terms of corrosion inhibition between the coatings. The ZrTPO-based film is very homogenous, thin and shows excellent adhesion to the magnesium chip. In contrast, the PTMOS-based film is much thicker, less homogeneous and some parts of it were removed. The combined coating was even less homogeneous than the PTMOS-based film.

In another approach, ZrO_2 nanoparticles are used as a reservoir for the storage and prolonged release of a corrosion inhibitor, such as cerium ions, for aluminium alloys. The resulting nanostructured doped sol–gel can be proposed as a potential candidate to substitute for the chromate pre-treatments of AA2024-T3.[16] The films are fabricated from TEOS and 3-glycidoxypropyltrimethoxysilane (GPTMS) precursors. The hybrid sol is doped with zirconia nanoparticles with cerium nitrate as corrosion inhibitor. Atomic force microscopy (Figure 4.13) shows that the nanosized particles are incorporated into the film matrix with a uniform distribution.

The hybrid ORMOSIL-based films in fact lose activity when the coating is partially destroyed. The nanoparticles not only reinforce the hybrid matrix but also absorb inhibitor ions releasing them during contact with moisture leading to effective self-healing properties. The prolonged release of inhibitor provides long-term corrosion protection for the aluminium alloy AA2024.

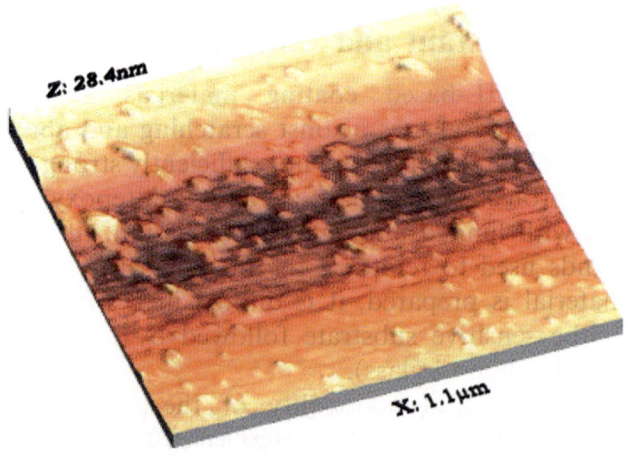

Figure 4.13   Atomic force microscopy topography of a hybrid sol–gel film. (Reproduced from ref. 21, with permission.)

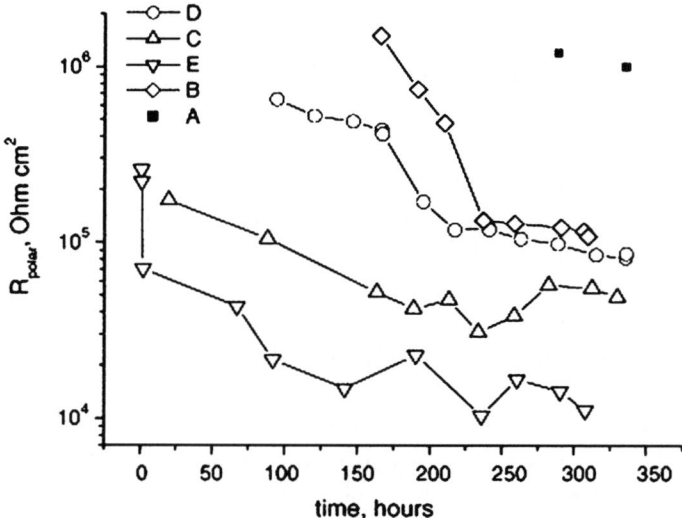

Figure 4.14 Evolution of polarization resistance for various coatings during immersion tests. (Reproduced from ref. 21, with permission.)

The evolution of polarization resistance (Figure 4.14) during immersion tests shows that AA2024-T3 coated with the doped coating (coating A) shows the first signs of corrosion processes only at the final stage of the immersion tests after 275 h of immersion. This is an order of magnitude longer than in the case of the other coatings.

4.5 Scratch-Resistant and Anti-ageing Coatings

Numerous ORMOSIL-based coating materials are commercially employed as protecting glasses against scratching and abrasion of sensitive surfaces. ORMOSIL alone are not sufficiently strong mechanically; therefore, these protective materials are based on (epoxy) alkoxysilanes polycondensed with limited amounts of aluminium, titanium or zirconium compounds used to reinforce the final structure. Once the polycondensed material is prepared, it is diluted in a specific solvent and deposited on the sensitive substrate followed by UV and/or thermal curing (UV-curable hybrid glass).

The composition and functionality of the systems greatly vary according to the substrate material to be coated, as they demand different curing conditions, adapted physical properties and characteristics as well as functionality for better wetting and adhesion. Dyes, pigments and fillers may also be added for specific applications, leading to smart multifunctional hybrid coatings. Protective coatings, for example, are supplied by TOP in Germany which manufactures lacquers developed by the Fraunhofer-Institut für Silicatforschung (ISC),[17] or by Hybrid Glass Technologies in the USA.

Coating formulations (lacquers) are clear, transparent, low-viscosity, solvent-free liquids that are stable with a shelf life of several months and are suitable for optical applications ranging from the ophthalmic to the optical fibre market. As an example of the former, Figure 4.15 shows a plate half coated with a layer of ABRASIL coating material only a few micrometres thick with the scratch traces from steel wool being observed on the uncoated half.[18]

One of the largest applications of inorganic–organic hybrid materials is in automotive coatings. For this technology, hybrid coatings are required that provide not only colouration but also scratch-resistance and protection from environmental factors such as UV and chemical attack. These requirements were firstly realized by DuPont with Generation 4®, which consist of a complex mixture of two hybrid polymers cross-linked simultaneously during curing to form a polymer network that is partially grafted and partially interpenetrated. An acrylate tetrapolymer core of high cross-link density, which includes methacryloxy-propyltrimethoxysilane and residual unsaturation, is generated. This gives a high-modulus and scratch-resistant function, dispersed in a polymer of low cross-link density, which provides film-forming properties. The superior scratch and environmental etch resistance of these coatings led to their acceptance as topcoats for eight of the ten top-selling automobiles in 1997.

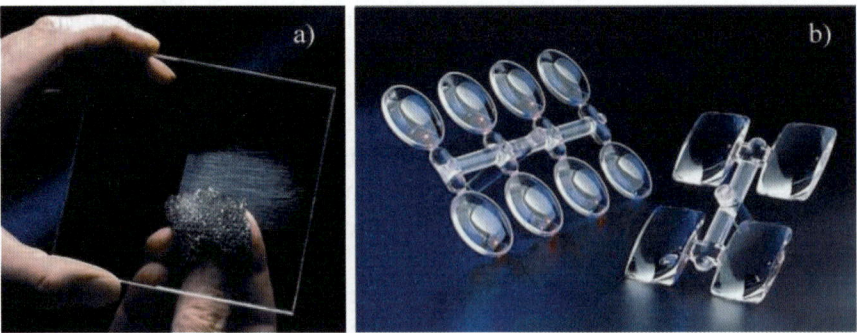

Figure 4.15   Plate half coated with a layer of ABRASIL coating material only a few micrometres thick. (Reproduced from ref. 18, with permission.)

In 2007 a new line of organofunctional and alkylsilane water-based sol–gel products called Dynasylan Hydrosil (trademarks of Evonik) entered the market.[19] They provide good corrosion and abrasion protection for valuable metal surfaces. They also provide coatings of excellent adhesion, without the need to use chromium(VI), a carcinogenic metal ion still widespread in corrosion treatment formulations.

The improvements provided by silanes in practical applications are based on multiple benefits (functions) offered by silanes (Figure 4.16). Silanes form permanent chemical bonds between substrates and organic polymers or resins, and their use in coating formulations results in excellent adhesion to challenging substrates such as glass, aluminium, steel and concrete. In challenging environments (bridges, ships, and other installations that are particularly susceptible to corrosion), aminosilanes are used to cross-link silicone-based epoxy resins with the formation of very hard and weather-resistant coating films. Finally, silane-modified inorganic pigments can be easily incorporated into coating formulations allowing, for example, hydrophilic pigments to be hydrophobized with alkylsilanes. The improved compatibility of the hydrophobized pigments helps achieve higher levels of filler content, lower viscosities, longer shelf life, better sedimentation resistance, improved mechanical properties of the cured coating and increased UV stability.

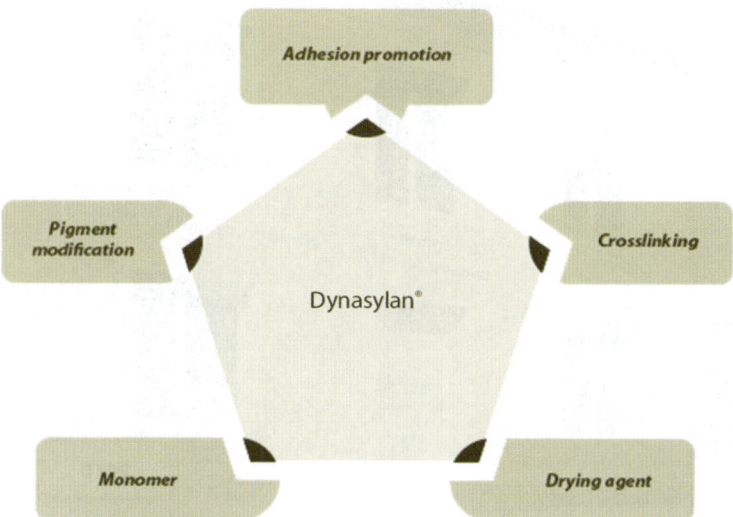

Figure 4.16 The improvements provided by silanes in practical applications are based on multiple benefits (functions) offered by silanes. (Reproduced from dynasylan.com.)

Applied as thin layers, which reduces material usage and weight, these coatings are especially important in the aerospace industry, where weight issues are an important factor. They can be used as primers or protective coatings for coil and sheet coatings, powder coatings, fasteners, automotive parts, finished sealers, steel furniture, appliances and other equipment.

The first ORMOSIL-based facade coating (Figure 4.17) on the market is a formulation that contains a dispersion of organic plastic polymer particles in which nanoscale particles of silica are incorporated and evenly distributed. Because of this combination of elastic organic material and hard mineral, coatings based on this novel nanobinder (called Col.9, BASF) combine in one product the advantages of conventional coating types in terms of low dirt pick-up, chalking and rack resistance, and colour retention (Figure 4.18).[20]

Comprehensive comparative tests involving traditional dispersion coatings have shown that facade paints containing Col.9 are durable. These findings are relevant because, especially in large conurbations, the high levels of carbon black and dust pollution rapidly lead to soiled building frontages and high renovation costs. The inorganic

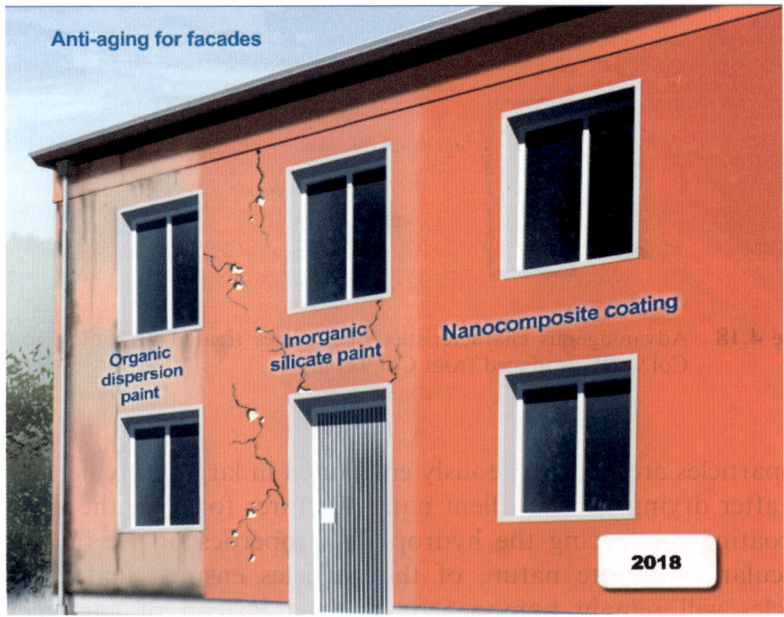

Figure 4.17   New nanobinder Col.9 combines in one product the advantages of conventional coating types in terms of low dirt pick-up, chalking and rack resistance, and colour retention. (Reproduced from Col.9.com)

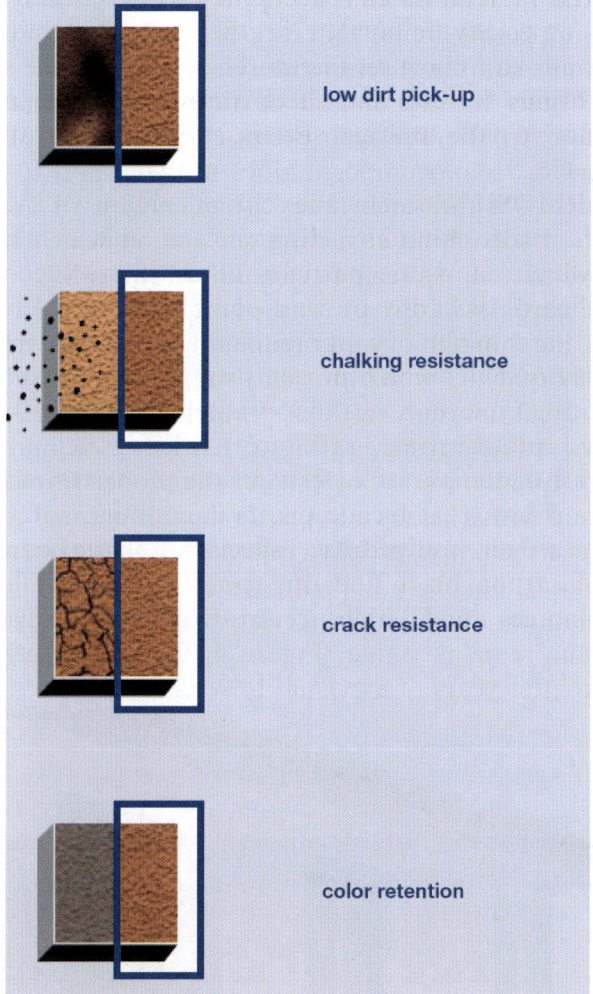

Figure 4.18 Advantageous characteristics of surfaces treated with the nanobinder Col.9. (Reproduced from Col.9.com)

nanoparticles are homogeneously embedded in larger polymer particles and, after drying, dirt-repellent nanostructures form on the surface of the coating reinforcing the hydrophilic properties of the facade. The molecular composite nature of the particles ensures that the nano-particles will remain homogeneously fixed instead of agglomerating when the liquid coating is formulated with water and colour pigments. This enables the formation of a stable three-dimensional network of nanoparticles on the surface which covers the entire film.

Unlike brittle, mineral-based coatings, the widely used synthetic resin-based dispersion paints are highly crack-resistant. But in summer, when dark house walls can reach temperatures of 80 °C in the sun, the synthetic resin begins to soften, and particles of soot and other contaminants stick to the surface. Because of its high silica content, however, the Col.9 nanocomposite does not succumb to this thermoplastic tackiness. At the same time, the mineral particles provide the coating with a hydrophilic attracting surface on which raindrops are immediately dispersed. As regards cleanliness, this offers a dual benefit: in heavy rain, particles of dirt are washed off extensively from the facade surface. Also, the thin film of water remaining when the rain has stopped dries extremely quickly, which prevents mould formation. In contrast, the rain rolling off unevenly in thick droplets from water-repellent surfaces of fully synthetic resin coatings often leaves behind unattractive streaks of dirt. Finally, while Col.9 shares the properties of hardness and low dirt pick-up with silicate coatings, its flexible organic content means it does not have their susceptibility to cracking that can rapidly lead to the entire coating flaking off. Permanently incorporating the mineral nanoparticles in the much larger acrylate polymer particles means that the colour tone remains stable (Figure 4.19) and there is no surface chalking even after years of exposure to weather.

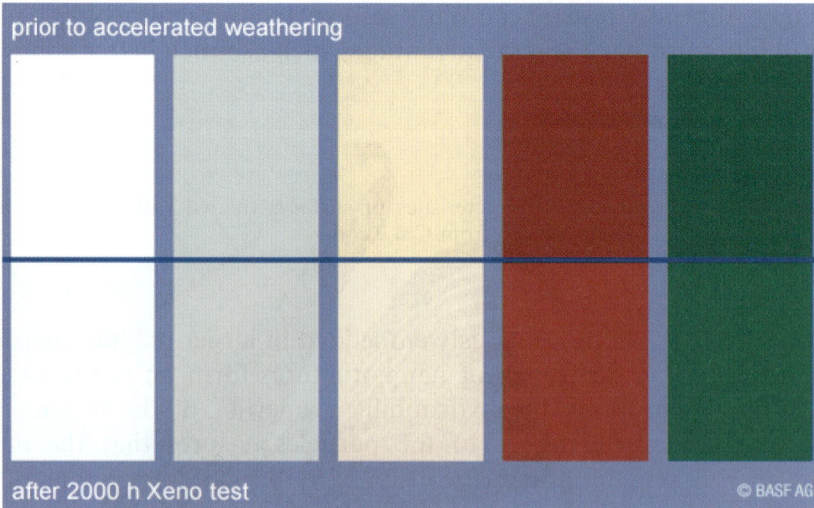

Figure 4.19 Colour retention of coloured surfaces treated with the nanobinder Col.9 from BASF. (Reproduced from Col.9.com)

4.6 Decorative Coatings and Barrier Systems

Decorative hybrid coatings are increasingly used on glass (Figure 4.20), mineral and metal surfaces. For example, a series of industrial coatings are based on an ethanol solution of various silica- or aluminosilica-based formulations (SC from Hybrid Glass Technologies), and are made available as clear liquids or coloured formulations in eight variations: clear, yellow, orange, pink, red, brown, blue and black.

Spin-, dip- or flow-coating techniques are used to apply these formulations on flat windows or sandblasted glass substrates. These coatings can be applied on glass, metallic or plastic surfaces to achieve hard 1–10 μm protective coatings and exhibit superior protective anti-staining properties and water repellency when compared to several existing commercial products. Indeed, they form strong chemical bonds with the glass surfaces and protect against staining or water corrosion.

Thin-film coatings made of dye-doped ORMOSIL are increasingly employed in industry as coloured decorative materials for glasses and plastics. There are more than 7000 types of organic dyes: all readily available, they have several advantages over pigments when they are used in coloured thin films because of their greater light absorption and transparency, non-scattering character and high solubility.[21]

Homogeneity of dispersion and the homogeneous nature of the hybrid ORMOSIL systems result in intensely coloured films even at low thickness. Compared to organic polymers, the coloured hybrid films have enhanced mechanical properties such as greater hardness and stronger abrasion resistance. With respect to inorganic sol–gel films, hybrid films also have several advantages such as much better adhesion,

Figure 4.20   Decorative hybrid coatings are increasingly used on glass surfaces.

transparency and flexibility, reduced surface roughness and better refractive index matching.

Films made of poly(methyl methacrylate) (PMMA)–silica hybrid doped with red, blue and green organic dyes and coated on glass substrates provide a clear example of the versatility of the sol–gel ORMOSIL technology. Here, in fact, the smooth polycondensation of a precursor solution made of TEOS–TMSPM–MMA (1:0.5:1 molar ratio of reactants; MMA: methyl methacrylate; TMSPM: 3-(trimethoxysilyl)propyl methacrylate) followed by radical polymerization of MMA ensures the formation of homogeneous and hard films having smooth surfaces with very little roughness (Figure 4.21).

The hardness as measured by modulus of all the hybrid films is at least 9H showing that the inorganic component in the hybrid films reinforces the organic component, producing harder films with more abrasion resistance as compared with pure PMMA films. The optical transparency of such hybrid films is another indication of their homogeneous composition of both organic and inorganic phases. Finally, the transparent coloured hybrid films have a very good optical quality, reflected in the interference oscillations (Figure 4.22), with colour intensity depending on the amount of organic dye in the films.

Hybrid materials are excellent sealant agents. Hence, for example, solar cells can be efficiently isolated with a vapour-deposited ORMOSIL layer (polycondensates of phenyl- and epoxy-functionalized alkoxysilanes mixed with smaller amounts of aluminium alkoxides) that ensures low cost and long-term durability of commercial photovoltaic modules (Figure 4.23) acting as a barrier against water vapour and gases, as well as an outside layer for protection against the weather. Furthermore, thin ORMOSIL films can be easily deposited in the roll-to-roll processes that are increasingly used to manufacture flexible photovoltaic modules at much lower cost compared to traditional silicon-based panels.[22]

Hydrophobic silane formulations such as those of the water-based Dynasylan line are excellent surface protection agents for wood (Figure 4.24), as the resulting coating repels *both* oil (dirt) and water, protects the surface from the effects of weathering and reduces the growth of microorganisms. The silane formulation does not affect the expansion and contraction behaviour of the wood and is therefore particularly suitable for outdoor objects without ground contact. The resulting coating has outstanding properties. The properties of silane-based wood coatings are based on covalent chemical reaction between the silanes and the hydroxyl groups at the wood surface (Figure 4.25). Long alkyl groups stable against UV-induced degradation ensure

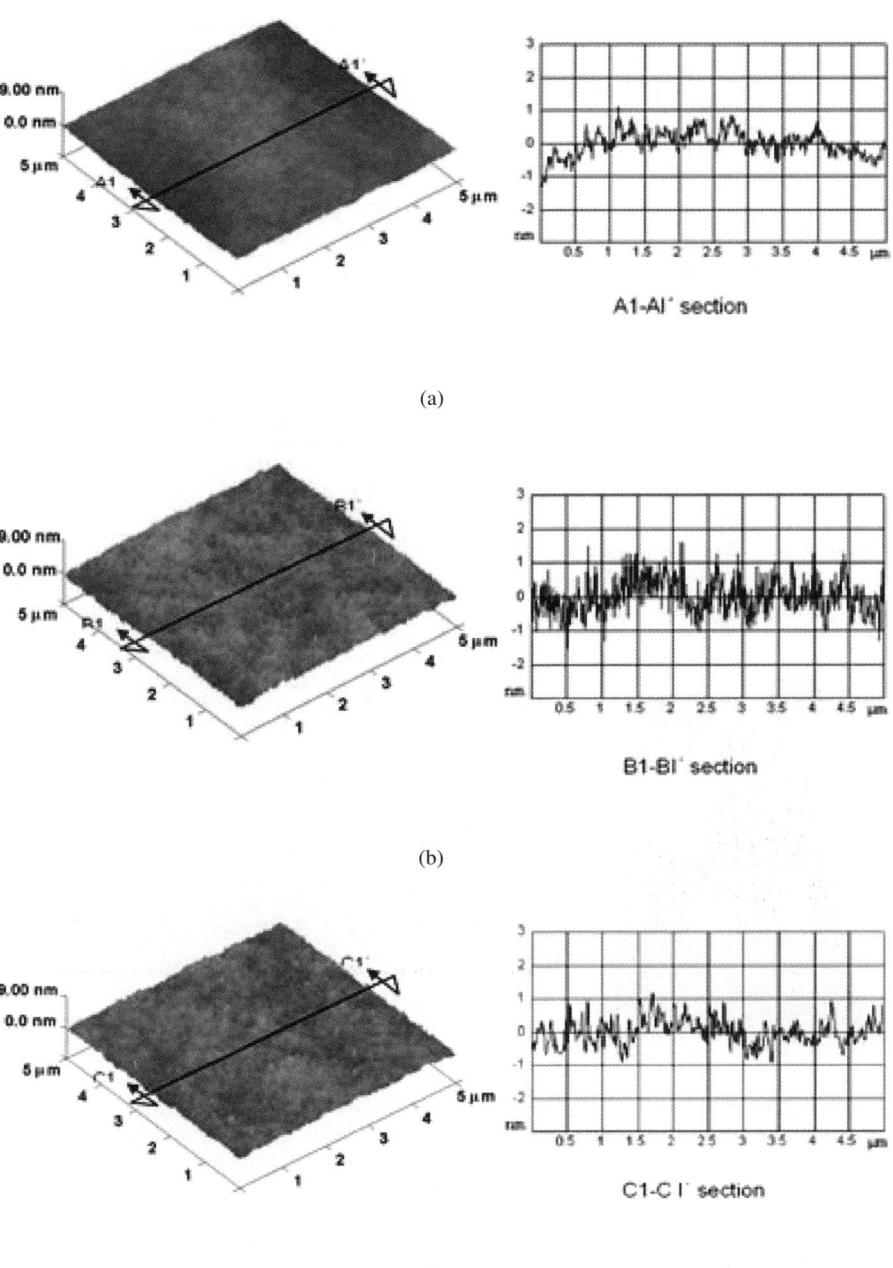

Figure 4.21 Atomic force microscopy images showing the morphology of coloured SiO$_2$–PMMA hybrid coatings with molar ratio formulation of 1 : 0.5 : 1.0 TEOS–TMSPM–MMA, with different concentrations and types of colour: (a) no colour; (b) 0.17 wt% of blue colour; (c) 0.83 wt% of green colour. The values of the r.m.s. average roughness measured for the three films was 0.47, 0.65 and 0.45 nm, respectively. (Reproduced from ref. 21, with permission.)

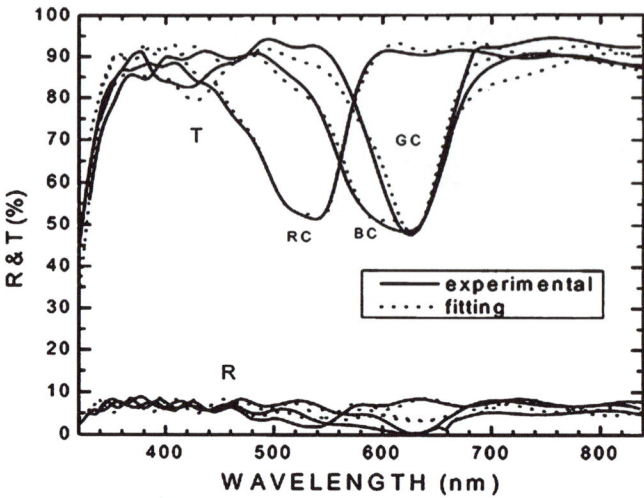

Figure 4.22 Transmittance (T) and reflectance (R) spectra of three coloured hybrid films. (Reproduced from ref. 21, with permission.)

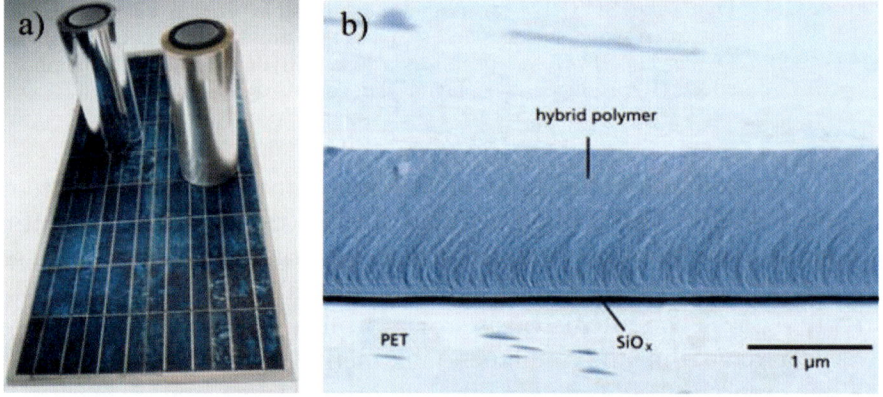

Figure 4.23 Solar cells are efficiently isolated with a vapour-deposited ORMOSIL layer that ensures long-term durability of commercial photovoltaic modules. (Reproduced from ref. 18, with permission.)

durability and provide hydrophobicity; oleophobicity is due to the overall siloxane structure.

With careful selection of the combination of starting alkoxides and appropriate synthesis conditions, silica-based sol–gel technology is now widely applied to the conservation of art objects and cultural heritage. For example, a hybrid sol–gel coating protects the fourteenth-century

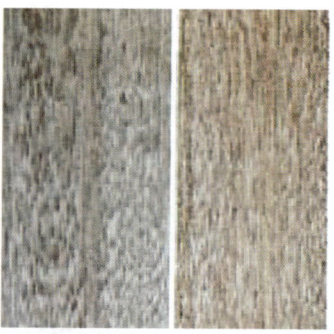

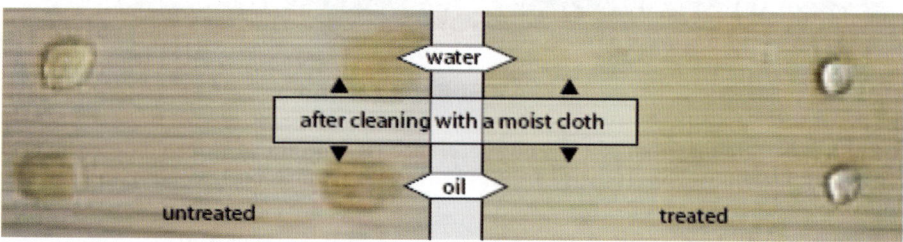

Figure 4.24   Dynasylan SIVO 121 forms an almost invisible protective layer (top). On untreated wood surfaces and surfaces that are weathered and varnished, Dynasylan SIVO 121 produces a strong hydrophobic and oleophobic effect. (Reproduced from Dynasylan.com)

mosaic situated above the gates of St Vitus cathedral, in the centre of Prague Castle.[23] The coating selected for the treatment of the entire mosaic is a multilayer system in which an organic–inorganic sol–gel layer is placed between a glass substrate and a fluoropolymer coating (Figure 4.26). The system consists of a protective coating (one sol–gel layer derived from epoxy silanes and methyl silanes, coated with a partially cross-linked functionalized fluoropolymer) and a sacrificial, removable coating consisting of the same fluoropolymer, not cross-linked, which is removed and reapplied periodically.

The sol–gel layers underneath the top polymer layer are estimated to last for 25 years. Beyond the long-term stability under severe ageing conditions compared to polymers, the advantages of such materials lie in their ability to adhere to many substrates, the possibility of making thicker coatings than with purely inorganic sol–gel systems and the ease of application using a brush. All organic polymers used in previous protection attempts have failed to stop corrosion, because of their poor durability, poor adhesion to glass and large diffusion coefficients for SO_2 and water. The mosaic itself is a large outdoor panel 13 m wide and 10 m

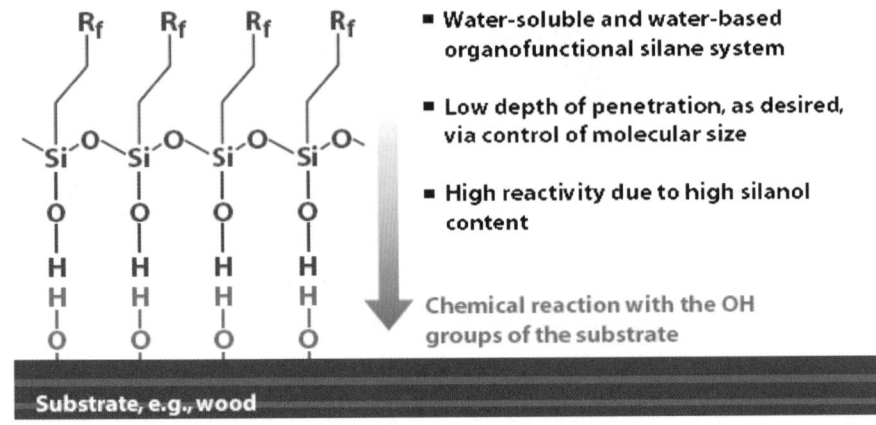

- Water-soluble and water-based organofunctional silane system

- Low depth of penetration, as desired, via control of molecular size

- High reactivity due to high silanol content

Chemical reaction with the OH groups of the substrate

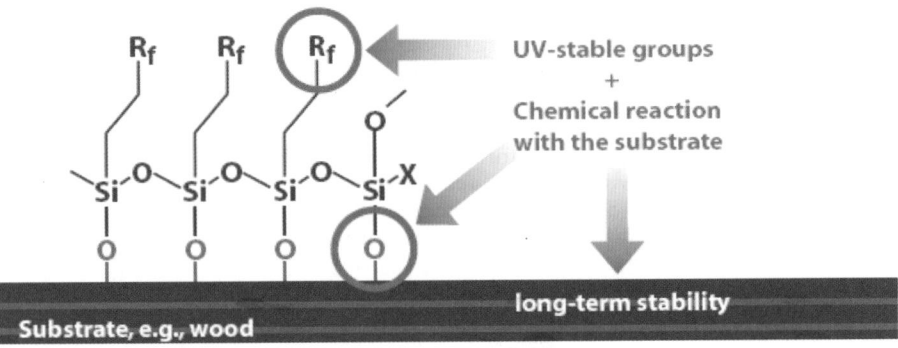

UV-stable groups
+
Chemical reaction with the substrate

long-term stability

Figure 4.25 Strong covalent bonding of silanes to the surface of wood ensures durable protection for all wood surfaces. Optimal treatment of $1\,m^2$ with $0.1\,L$ Dynasilan ensures three years' protection. (Reproduced from Dynasylan.com)

high, made from about a million pieces of multicoloured, high-potassium glass embedded in a mortar. The glass is chemically unstable as it is exposed to harsh weather conditions (high sulfur dioxide levels, rain and temperature varying between -30 and $+65\,°C$), in which the alkali reacts with the atmosphere and water to form salt deposits. The sol–gel coating has the same consistency as paint and can be applied with a small brush by hand in order to avoid coating the interstitial space between tesserae. This is followed by curing at $90\,°C$ for two hours using large infrared lamps.

Similarly, GPTMS-based compositions of a solution of pre-hydrolysed GPTMS–methyltrimethoxysilane (MTMS) hybrid doped with colloidal

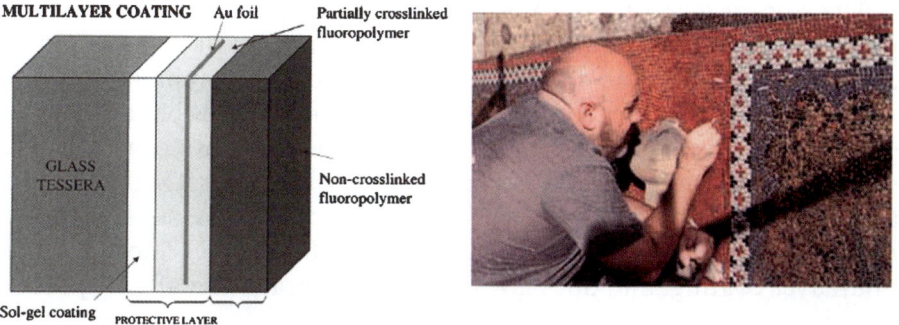

Figure 4.26 Schematic cross-section (left) of a multilayer coating used in the long-term protection of a mosaic. The mosaic was originally gilded; therefore, a tiny gold foil is inserted into each tessera. Professor E. Bescher (right) brushing the mosaic with the sol–gel paint. (Photo courtesy of UCLA Daily Bruin). (Reproduced from ref. 23, with permission.)

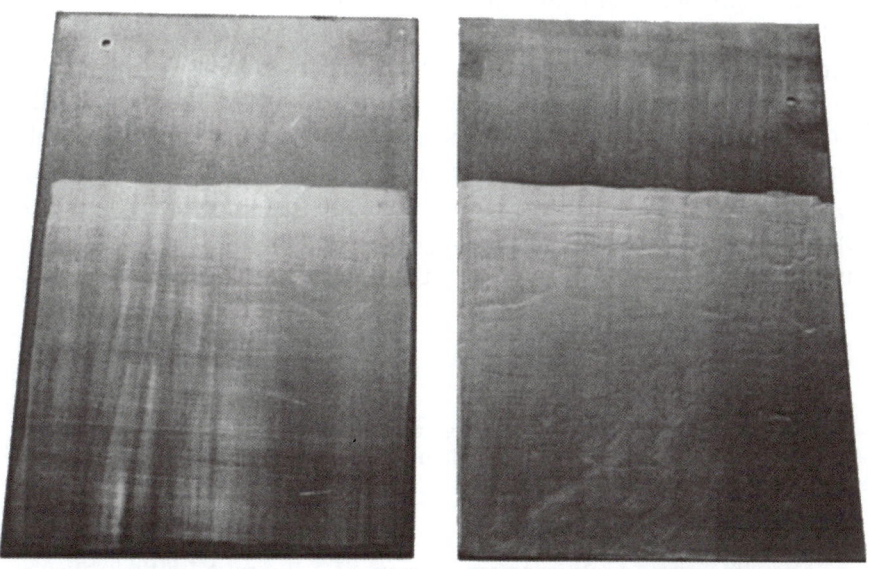

Figure 4.27 Brass samples after two years' exposure in a high-sulfur, high-humidity environment. The left plate is coated with GPTMS–MTMS; the right plate further coated with a Lumiflon polymer. (Reproduced from ref. 24, with permission.)

silica particles (colloidal SiO_2 improves the mechanical properties while maintaining good optical transparency) brushed over metallic art objects, such as brass or bronze, subject to corrosive environments effectively delay corrosion. These compositions exhibit good resistance to UV light.[24]

The appearance of GPTMS–MTMS and GPTMS–MTMS–fluoro-polymer coatings on brass plates after several years of exposure in an ageing chamber (Figure 4.27) shows evident corrosion inhibition due to the sol–gel layer; the top uncoated portion of the plates exhibits extensive corrosion. In the bottom part of the left plate coated with a single layer of the ORMOSIL solution onset of deterioration had begun along the edges of the plate, probably due to excess stress in these regions. The plate on the right was coated with a sol–gel/fluoropolymer multilayer coating, and does not shown any degradation after exposure.

4.7   Functionalisation of Textiles by Sol–Gel Coatings

The maintenance and improvement of current properties and the creation of new material properties are the important reasons for the functionalisation of textiles with sol-gel organosilica coatings with particle diameters smaller than 50 nm (nanosols).[25] Coating with nanosols enables the manifold alteration of their physico-mechanical, optical, electrical and biological properties of a textile surface (Figure 4.28).

Within today's global textile market worth more than $400 billion high-grade textiles rapidly grow as functional textiles improve applications (enhanced comfort, easy care, health and hygiene and extend their range of utilisation (ensuring protection against mechanical, thermal, chemical and biological attacks) affording technical textiles, with numerous usages in automotive, railroad and aviation engineering, in construction and for home textiles.

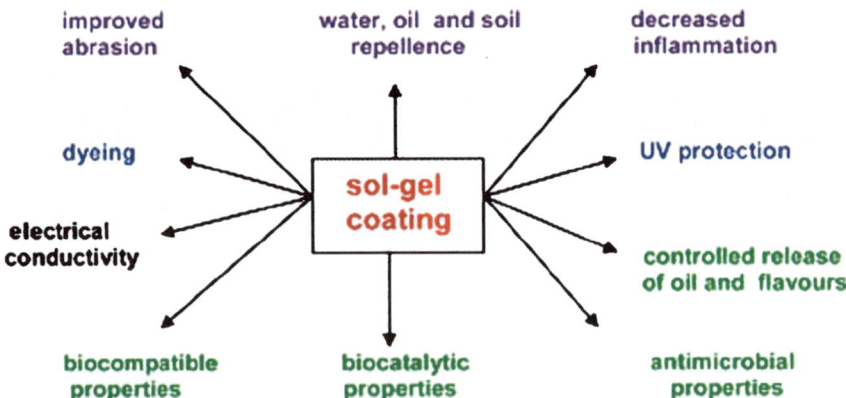

Figure 4.28   Some possibilities of surface functionalisation by modified nanosols. (Reproduced from ref. 25, with permission.)

New products that are being developed include textiles with water, oil and soil repellency and with antimicrobial properties. In general, the deposition makes use of a prehydrolyzed nanosol obtained by hydrolytic condensation of organosilanes (Scheme 4.2).

Coating using the nanosol can either be performed by using the alcogel resulting from the hydrolysis step or, better from an environmental viewpoint, by employing the *hydrogel* easily produced by passing air through the alcogel and simultanously substituting water.[26] The physical and chemical properties of the coating are widely tailored either by changing the non-hydrolyzable moiety, or by doping the nanosol with a suitable species. In any case, the nanoparticulate size of the sol particles promotes excellent adhesion to the textile fibres, which can be further enhanced by subsequent thermal treatment (Figure 4.29).

For example, sol-gel immobilised bioactive liquids such as cineol, camphor, menthol, evening primrose and perilla oil used to functionalise textiles afford either skin-friendly textiles with antimicrobial and anti-allergic effects due to immobilised natural oils; or textiles for therapeutic treatment of the respiratory tract by means of immobilised mixtures of high volatility natural agents such as eucalyptol, camphor and menthol.[27]

In another example, by depositing a continuous film of the nanosol based on long-chained alkylsilane additives onto the textile surface

$$(OR)_3SiR' \xrightarrow{\text{hydrolysis}} R'(SiO_{1.5})_n \xrightarrow{\text{coating}} R'(SiO_{1.5})_m \xrightarrow{\text{drying}} R'(SiO_{1.5})_m$$

precursor (I)   *nanosol (II)*   *lyogel film (III)*   *xerogel film (IV)*

Scheme 4.2 Preparation of sol-gel nanocoatings (Adapted from ref. 25, with permission.)

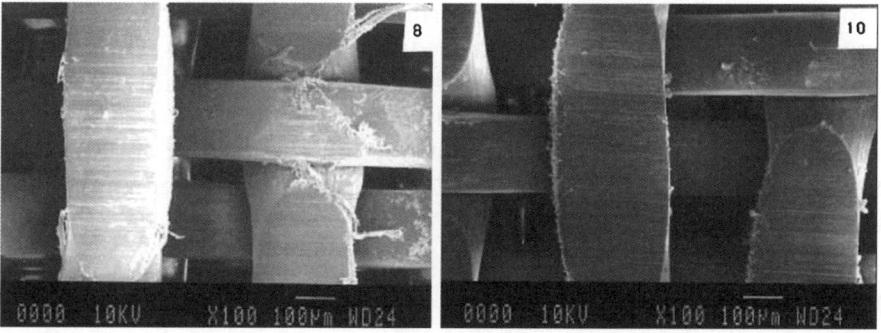

Figure 4.29 SEM pictures of polyester sieves after abrasion without (*left*) and with nanosol coating (*right*). (Reproduced from ref. 25, with permission.)

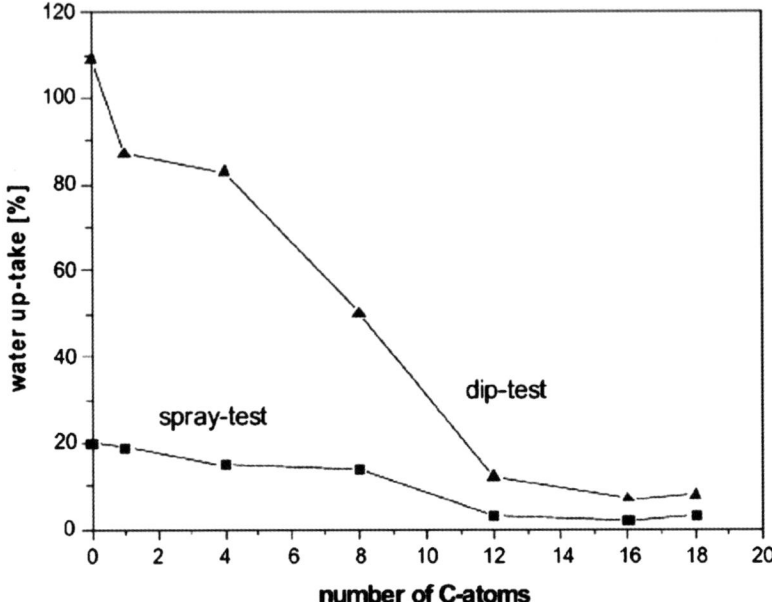

Figure 4.30 Water up-take of CO/PES fabrics with silica coatings containing alkyl-silane additives with increasing alkyl chain length after spraying with or dipping in water (for uncoated fabric the water uptake was 99% after spraying and 157% after dipping). (Reproduced from ref. 28, with permission.)

results in excellent water-repellent coatings.[28] For, an alkyl chain containing more than at least 12 carbon atoms water up-take in a spray-test is reduced to as low as 20% with very small concentrations of long-chained alkylsilanes (Figure 4.30).

By using these non-fluorinated coatings on textiles, thus, hydrophobicity similar to that of fluorinated coatings is possible while retaining stability to washing procedures. Finally, the very low layer thickness (below 1 μm) that makes the consumption of coating solutions very low. Overall, the use of the sol-gel nanotechnology overcomes the limitations of applying conventional methods to impart certain properties to textile materials and sol-gel coated textiles is being applied into several area of textile industry.[29]

References

1. B. Arkles, Commercial applications of sol–gel-derived hybrid materials, *MRS Bull.*, 2001, May, 402.

2. G. Palmisano, E. Le Bourhis, R. Ciriminna, D. Tranchida and M. Pagliaro, ORMOSIL thin films: tuning the mechanical properties via a nanochemistry approach, *Langmuir*, 2006, **22**, 11158.
3. Fluoropolymer-containing sol-gel coating, WO/2004/076570, 2004.
4. M. Kursawe, R. Anselmann, V. Hilarius and G. Pfaff, Nanoparticles by wet chemical procession in commercial applications, *J. Sol-Gel Sci. Technol.*, 2005, **33**, 71.
5. M. Zayat, P. Garcia-Parejo and D. Levy, Preventing UV-light damage of light sensitive materials using a highly protective UV-absorbing coating, *Chem. Soc. Rev.*, 2007, **36**, 1270.
6. P. Garcia-Parejo, M. Zayat and D. Levy, Highly efficient UV-absorbing thin-film coatings for protection of organic materials against photodegradation, *J. Mater. Chem.*, 2006, **16**, 2165.
7. Y. Tang, J. A. Finlay, G. L. Kowalke, A. E. Meyer, F. V. Bright, M. E. Callow, J. A. Callow, D. E. Wendt and M. R. Detty, Hybrid xerogel films as novel coatings for antifouling and fouling release, *Biofouling*, 2005, **21**, 59.
8. M. R. Detty, M. D. Drake, Y. Tang and F. V. Bright, Hybrid antifouling coating compositions and methods for preventing the fouling of surfaces subjected to a marine environment, *US Pat.*, 7 244 295 B2, 2007.
9. P. Selvaggio, M. Pagliaro, R. Ciriminna, S. Tusa, M. R. Detty and F. V. Bright, Ecofriendly protection from biofouling of the monitoring system at Pantelleria's Cala Gadir Archaeological site, *J. Nautical Archael.*, 2009, **38**, xx.
10. See: www.safenanotech.com.
11. L. Mascia and T. Tang, Ceramers based on crosslinked epoxy resins–silica hybrids: low surface energy systems, *J. Sol-Gel Sci. Technol.*, 1998, **13**, 405.
12. L. Mascia, L. Prezzi, G. D. Wilcox and M. Lavorgna, Molybdate doping of networks in epoxy–silica hybrids: domain structuring and corrosion inhibition, *Prog. Org. Coat.*, 2006, **56**, 13.
13. D. Mandler and Y. Tamar, Corrosion inhibition of magnesium by combined zirconia silica sol-gel films, *Electrochim. Acta*, 2008, **53**, 5118.
14. A. N. Khramov, V. N. Balbyshev, L. S. Kasten and R. A. Mantz, Sol–gel coatings with phosphonate functionalities for surface modification of magnesium alloys, *Thin Solid Films*, 2006, **514**, 174.
15. A. L. K. Tan, A. M. Soutar, I. F. Annergren and Y. N. Liu, Multilayer sol–gel coatings for corrosion protection of magnesium, *Surf. Coat. Technol.*, 2005, **198**, 478.
16. M. L. Zheludkevich, R. Serra, M. F. Montemor, K. A. Yasakau, I. M. Miranda Salvado and M. G. S. Ferreira, Nanostructured sol–gel

coatings doped with cerium nitrate as pre-treatments for AA2024-T3 corrosion protection performance, *Electrochim. Acta*, 2005, **51**, 208.

17. A research institute with a long tradition in the development of industrial inorganic and hybrid coatings, which owns the trademark ORMOCER. See: www.ormocer.de.
18. C. Sanchez, B. Julán, P. Belleville and M. Popall, Applications of hybrid organic-inorganic nanocompacts, *J. Mater. Chem.*, 2005, **15**, 3543.
19. See: www.dynasylan.com.
20. For an interesting video account on Col.9 and its action, see the movie online at: www.corporate.basf.com/en/stories/wipo/col-9.
21. J. L. Almaral-Sanchez, E. Rubio, J. A. Calderón-Guillén, A. Mendoza-Galvan, J. F. Pérez-Robles and R. Ramírez-Bon, Colored transparent organic–inorganic hybrid coatings, *AZojomo*, 2006, 2. (*Adv. Technol. Mater. Mater. Process.*, 2005, **7**, 203.).
22. M. Pagliaro, G. Palmisano and R. Ciriminna, *Flexible Solar Cells*, Wiley-VCH, Weinheim, 2008.
23. E. Bescher, F. Piqué, D. Stulik and J. D. Mackenzie, Long-term protection of the Last Judgment mosaic in Prague, *J. Sol-Gel Sci. Technol.*, 2000, **19**, 215.
24. E. Bescher and J. D. Mackenzie, Sol-gel coatings for the protection of brass and bronze, *J. Sol-Gel Sci. Technol.*, 2003, **26**, 1223.
25. B. Mahltig, H. Haufe and H. Böttcher, Functionalisation of textiles by inorganic sol-gel coatings, *J. Mater. Chem.*, 2005, **15**, 4385.
26. U. Soltmann, J. Raff, S. Selenska-Pobell, S. Matys, W. Pompe and H. Böttcher, *J. Sol–Gel Sci. Technol.*, 2003, **26**, 1209.
27. H. Haufe, K. Muschter, J. Siegert and H. Böttcher, Bioactive textiles by sol–gel immobilised natural active agent. *J. Sol–Gel Sci. Technol.*, 2008, **45**, 97.
28. B. Mahltig and H. Böttcher, Modified Silica Sol Coatings for Water-Repellent Textiles, *J. Sol–Gel Sci. Technol.*, 2003, **27**, 43.
29. Y. W. H. Wong, C. W. M. Yuen, M. Y. S. Leung, S. K. A. Ku and H. L. I. Lam, Selected applications of nanotechnology in textiles, *AUTEX Res. J.*, 2006, **6**(1), 1.

CHAPTER 5
Catalysis

5.1 Catalysis by Sol–Gels: An Advanced Technology for Organic Chemistry

The pharmaceutical industry produces between 25 and 100 kg or more of waste for every kilogram of active pharmaceutical ingredient (API) manufactured.[1] According to a leading practitioner of the industry, "the potential waste coproduced with APIs is in the range of 500 million to 2 billion kg per year. Even at a nominal disposal cost of $1 per kg, the potential savings just in waste avoidance is significant faced to the pharmaceutical industry annual sales (almost $500 billion in 2003)."[2]

Catalysis over heterogenized catalysts enabling *one-pot*, multistep synthesis[3] would offer a solution to most of these problems (Scheme 5.1) eliminating the large amounts of solvents and purification media currently employed by the industry. Yet "many homogeneous catalytic systems cannot be commercialized because of difficulties associated with separating the products from the catalyst".[4] Until recently, indeed, heterogeneous catalysts in the fine chemicals industry were remarkable for a disappointingly poor achievement. As of 2003 "to my knowledge—as put by Cole-Hamilton—the only commercial example of a homogeneous catalyst heterogenized on a solid support is the carbonylation of methanol using $[RhI_2(CO)_2]^-$ electrostatically bound to an ion exchange resin";[4] and even in that single case, the author pointed out that leaching could not be avoided.

The reason for this is mostly rooted in the fact that historically the fine chemicals industry is a product (and not process) oriented industry, *i.e.* it focuses on the development of new products to maximize revenues in the

Silica-Based Materials for Advanced Chemical Applications
By Mario Pagliaro
© Mario Pagliaro 2009
Published by the Royal Society of Chemistry, www.rsc.org

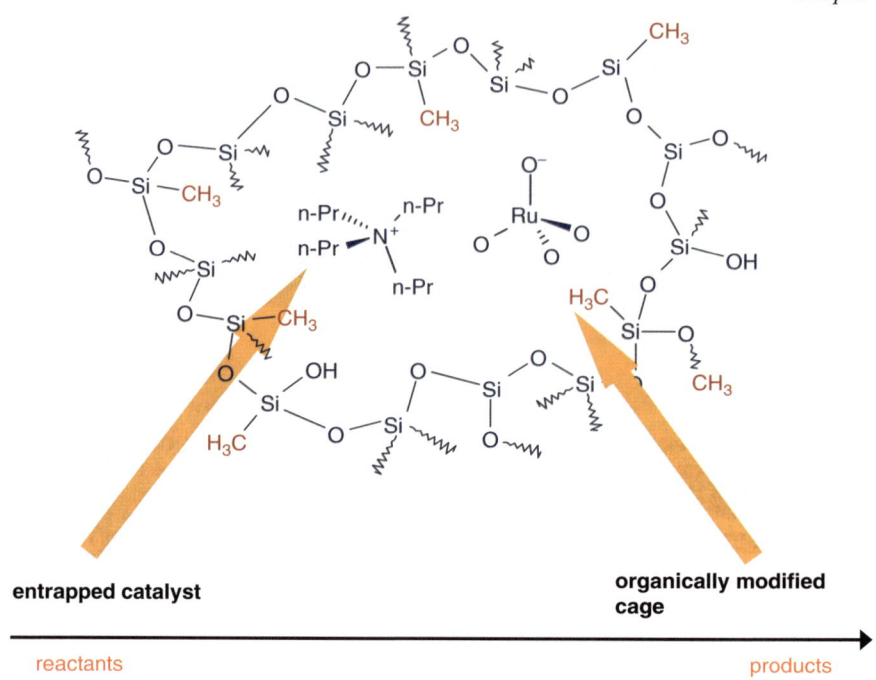

Scheme 5.1 A systems approach to organic synthesis with doped sol–gel nano-composites. (Reproduced from ref. 8a, with permission.)

short time span in which exclusive rights are granted by patenting an innovation. As a result, traditional heterogenization technologies of homo-geneous catalysts were characterized by a poor level of performance in terms of activity, selectivity and stability. Current economic hypercom-petition, however, and ever stricter environmental regulations are bringing about a radical change with industry (often small, knowledge-intensive companies) and academe working together to develop a variety of solid-phase catalytic technologies for high-throughput organic synthesis.[5]

Some of the most remarkable achievements include: microencapsula-tion in polystyrenes such as entrapped OsO_4 for olefin hydroxylation (exploiting the interaction between π-electrons of benzene rings of the polystyrenes used as polymer backbones and the vacant orbitals of the catalysts);[5] polyurea-entrapped palladium (PdEnCat)[6] for a multiplicity of C–C forming reactions; and the use of carboxylic acid-functionalized polymer (FibreCat).[7] In general, however, metal leaching cannot be avoided. The PdEnCat catalyst, for instance, leaches some 4% of pal-ladium per catalytic reaction run.

Along with these recent technologies, sol–gel entrapped catalysts made of organically modified silicates (ORMOSIL) doped with one or

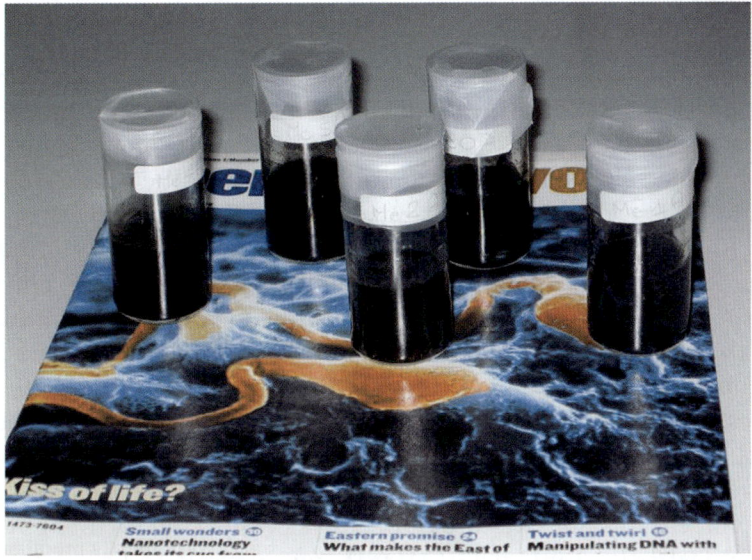

Figure 5.1 The alcogels shown are ORMOSIL doped with the ruthenium species tetra-*n*-propylammonium perruthenate (TPAP). Upon a mild heat treatment these materials become more active than TPAP in solution.

more catalytic species (Figure 5.1) enable heterogeneous conversions that are *more* selective and active than conventional homogeneous catalyses.[8] A number of recent studies have proved that catalysts encapsulated in the inner porosity of ORMOSIL according to the reaction

$$CH_3Si(OCH_3)_3 + Si(OCH_3)_4 + catalyst \xrightarrow[H_2O]{MeOH} catalyst@[(CH_3SiO_nH_m]_p \quad (5.1)$$

are, in fact, much more efficient than other supported catalysts and even more efficient than homogeneous catalysts in a wide variety of reactions, under largely different conditions.[8a]

The trend was studied and verified, for instance, for reactions catalysed by transition metal, organo- and enzyme catalysts entrapped in ORMO-SIL prepared by copolymerization of tetramethoxysilane (TMOS) and the modifying co-precursor methyltrimethoxysilane (MTMS). It has been correlated with the encapsulation itself but also with the structure of the sol–gel matrix, namely the hydrophobicity–lipophilicity balance (HLB) and the textural properties of the materials.[9]

Furthermore, the use of silica-based ceramic matrices as functional supports has a number of practical, unique advantages including stability towards harsh conditions, low swelling and consistent binding sites for the catalyst. In a recent review[2] it was emphasized how "inorganic supports, as a rule, lack the synthetic *flexibility* of organic shells. This deficiency may limit their broad use." Enabling one to overcome this limitation, ORMOSIL afford ideal integration between the excellent physical and chemical stability of silica, along with the immense versatility provided by the organic modification of the inorganic silica structure.[10] The outcome is a similarly vast range of new chemistries whose first commercial examples were numerous already in 2005. In practice, pharmaceutical companies can now rely on sol–gel entrapped catalysts for both rapidly generating new drug candidates, especially in high-throughput microreactors (Figure 5.2), and also—because of the MetaChip (see below)—to screen these candidates for toxicity. However, catalysis by sol–gel entrapped catalysts is powerful chemical technology whose potential, in terms of benefits to society, is far from being realized.

Our current, broader picture of the physicochemical factors governing the chemical behaviour of ORMOSIL reveals that chemical modification of the sol–cage does indeed alter the chemical properties of the entrapped dopant.[11] Learning to master this interaction for different

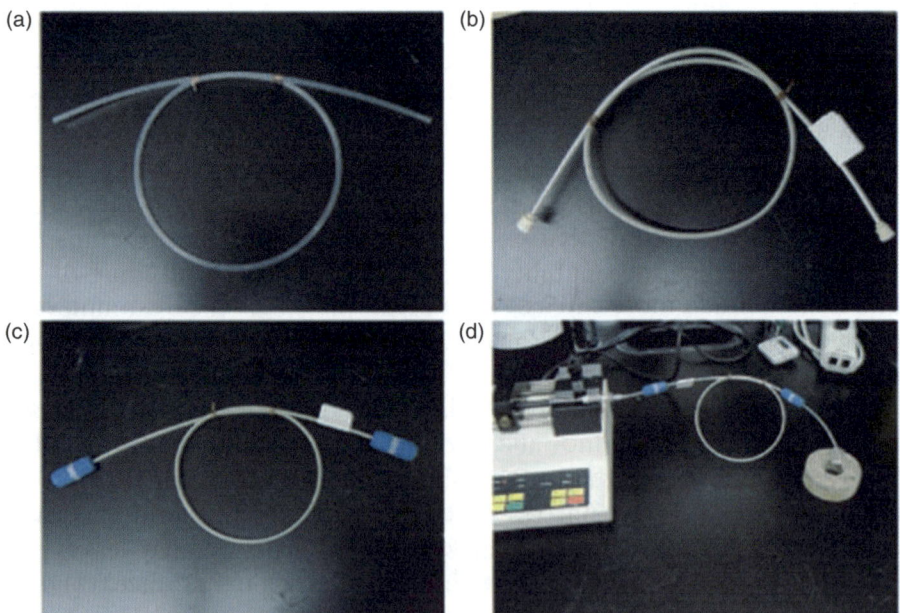

Figure 5.2 A simplified microreactor: (a) empty tubing (b) filled with AO resin with (c) filter caps and (d) attached to syringe pumps. (Reproduced from ref. 13, with permission.)

enzymes might thus lead to replication of the outstanding results obtained with entrapped lipase for which the schematic view of non-covalent interactions between the gel matrix and the lipase could well require revision. In conclusion, sol–gel ORMOSIL-entrapped catalysts are emerging as an advantageous technology capable of providing the fine chemicals and pharmaceutical industries with a number of efficient new solutions to face the challenges posed by the enhanced pace of innovation *and* sustainability, and eventually to make chemistry not only environmentally benign but also more efficient.

5.2 Sol–Gel Entrapped Catalysts: Tailored Catalytic Materials

Catalysis by sol–gel doped silica-based materials has become in the last 20 years a prominent tool to synthesize a vast number of useful molecules both in the laboratory and in industrial plants.[12] The underlying basic concept of all sol–gel applications is unique: one or more host molecules are entrapped by a sol–gel process within the cages of an amorphous metal oxide where they are accessible to diffusible reactants through the inner pore network, which leads to chemical interactions and reactions (Figure 5.3).

Several methodologies have been developed to entrap organometallic catalysts in solid materials including ship-in-a-bottle, grafting and anchoring techniques. In general all these traditional methods are surface heterogenizations in which one organic or inorganic polymer is mixed with a solution of the catalyst (or a precursor). Results are varied, but reduced catalytic activity, slower reaction rates due to transport limitations, lack of accessibility of the active sites and leaching of the supported species are commonly observed problems. For example, the powerful and versatile PdEnCat systems behave as heterogeneous sources for soluble, catalytically active species during the course of Heck and Suzuki couplings (Table 5.1).[13]

Indeed, one of the main concepts of heterogeneous catalysis—which is crucial to all applications of solid catalysts to organic chemistry—is that while derivatizations of material surfaces require the formation of a new covalent bond through a slow heterogeneous reaction that leaves the anchored molecules unprotected at the material's pore surface, *confinement* of an active species within a material's pores restricts the possible orientations that reactants can assume approaching the catalytic centre so that, in general, sol–gel entrapped catalysts show *higher* (enantio)selectivity in comparison both to homogeneous catalysts in solution or

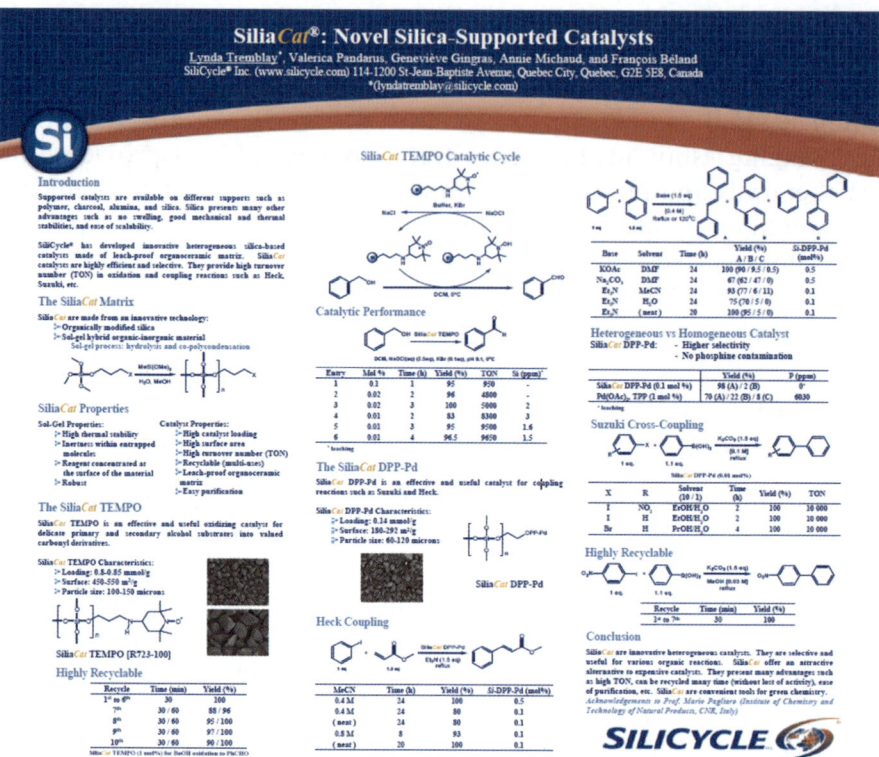

Figure 5.3 The SiliaCat line of organosilica-entrapped catalysts commercialized by SiliCycle is the first comprehensive product line addressing a variety of catalytic reactions.

Table 5.1 Inductively coupled plasma analysis of Heck reaction filtrates. (Reproduced from ref. 13, with permission.)

Catalyst	Solvent	Pd leached (%)
PdEnCat 30	Toluene	37
PdEnCat 30	Isopropyl alcohol	0.5
PdEnCat 30	Dimethylformamide	46
PdEnCat TOTP30	Toluene	42
PdEnCat TOTP30	Isopropyl alcohol	9
PdEnCat TOTP30	Dimethylformamide	52

surface-bound to a nonporous material.[14] Doped sol–gel silicas in fact are chromatographic materials which adsorb and concentrate the reagents at the cage surfaces where reactions take place while the adsorbed molecules promote further diffusion of incoming reactants.[15]

The chemical behaviour of the resulting catalyst is affected by four main factors:

1. The HLB.
2. The flexibility of the sol–gel cages.
3. The accessibility of the entrapped species.
4. The presence of entrapped co-dopants.

Each of these properties can be subtly controlled and tailored over a vast range of values enabling the preparation of catalysts capable of meeting the demanding requirements in terms of selective activity, stability and versatility (Figure 5.4).

In general, sol–gel catalysts are heterogeneous materials employing solid–liquid or solid–gaseous interphases. A mobile and a stationary component penetrate each other at the molecular level with the catalytic species being well-defined, highly *mobile* and *homogeneously* distributed

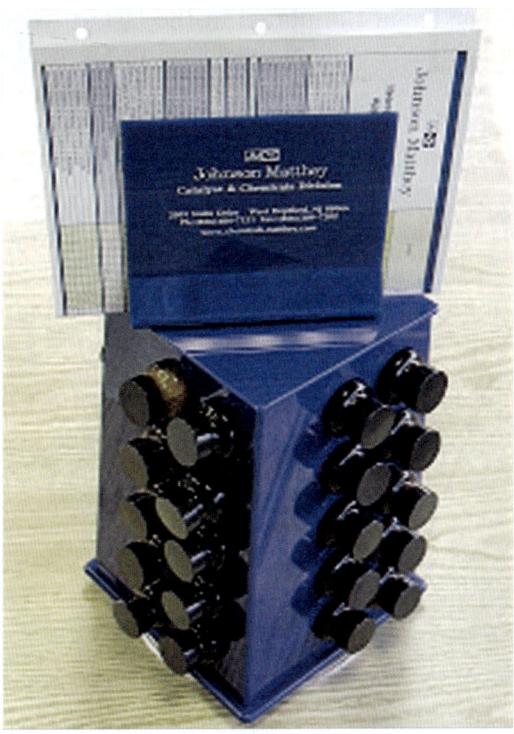

Figure 5.4   Requirements for heterogeneous catalysts for fine chemicals are often demanding. And this 40-sample screening kit offers instant availability to catalysts with varying characteristics and metal distributions. (Photo courtesy of Johnson Matthey.)

across a highly porous chemically and thermally inert network, thus combining the advantages of homogeneous (high selective activity and reproducibility) and heterogeneous (stabilization and easy separation and recovery of the catalyst) catalysis. Two methods are used to prepare sol–gel entrapped catalysts:

- physical entrapment of organometallic species (including enzymes); and
- chemical entrapment through polymerization of trialkoxysilyl derivatives of metal ligands.

In many cases, however, the silicas are formed under kinetic control,[16] and not thermodynamic, resulting often in "living" materials that undergo structural modifications, and thus changes in reactivity, even months after preparation (Figure 5.5).

Catalyst textural properties, HLB, shape and composition can thus be planned by a rational choice of the preparation conditions based on the fundamental knowledge accumulated in the last 20 years. For example, the protecting properties of the sol–gel caging are well established. For example, while fresh $RuCl_2(PPh_3)_3$ exposed to air gradually loses its high efficiency in mediating the double-bond migration in allylbenzene, the entrapped version can be used many times without any loss in

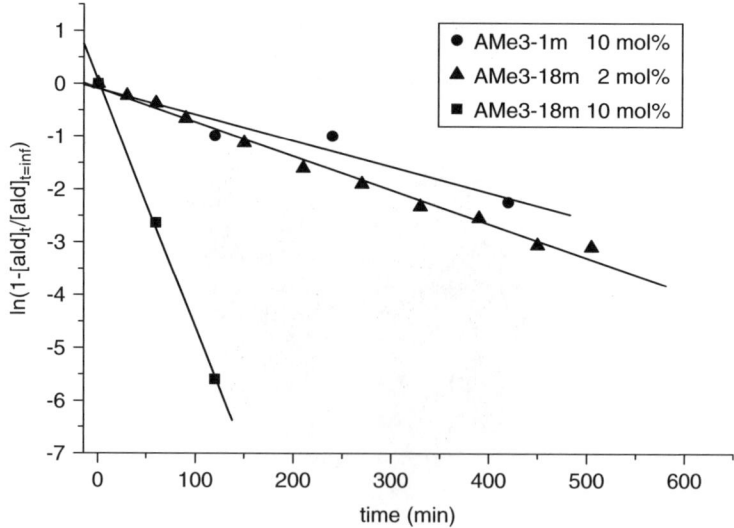

Figure 5.5   Example of drastic activity enhancements. Aerobic oxidation of benzyl alcohol to benzaldehyde in $scCO_2$ over TPAP entrapped in aged (■, ▲) and fresh (●) 75% methyl-modified silica matrix. (Reproduced from *Adv. Funct. Mater.*, with permission.)

Figure 5.6 Scanning electron microscopy images of a silica-entrapped palladium catalyst amenable for a variety of C–C forming reactions (particle sizes are from 60 to 125 μm).

activity.[1] High acidity and low water/silane ratio were thought to be required for achieving such a balance;[15] a closer look reveals that what is actually demanded is a thorough hydrolysis of the alkoxides in order to ensure an open cross-linked final network (Figure 5.6) in which the active molecules are entrapped at the cage surface, and not buried in the bulk of the material.[17]

During the growth of the material, in the monomer → oligomer → sol → gel → xerogel transition, micelles tend to form which may segregate the (generally hydrophobic) organometallic catalyst within the core and the polar head at the interface of the growing material. In order therefore to disrupt any intermediate micelles and to ensure homogeneous dispersion of the catalyst across the polymeric network, a large amount of co-solvent and high water/silane ratio are used to afford highly active entrapped catalysts.

Indeed, the *mobility* of the entrapped dopant is crucial in promoting the reactivity of the final materials. Thus, provided that the dopant molecules are at the surface and enjoy enough freedom, high porosity will certainly promote reactivity by limiting intraparticle diffusion; but that will *not* be the case if microporous xerogels of different HLB are compared (*cf.* entrapped lipase and tetra-*n*-propylammonium perruthenate (TPAP) where ORMOSIL with the smaller pores are *more* reactive).

5.3 Physically Entrapped Catalysts

In the 1990s a large variety of organometallic catalysts were physically entrapped in leach-proof silica gels showing enhanced activity and opening new possibilities.[1] A general finding emerging from these researches is that *only* the physically entrapped quaternary onium ion pairs of transition metal catalysts do not leach in solution during catalysis. Clearly, when the entrapped catalyst is insoluble in the reaction solvent, its encapsulation enables smooth, leach-proof catalysis. For instance enantioselective hydrogenation mediated by sol–gel entrapped Ru-BINAP can now be carried out in water where Ru-BINAP is entirely insoluble, and the catalyst can be reused at length.[1]

Physically doped ORMOSIL opened the route to the use of *water* as a reaction medium for catalytic organic reactions by simply emulsifying the organic reactants (all hydrophobic). In this manner, a general reaction procedure using a rhodium-doped 10–20% alkylated silica matrix in an elegant three-phase emulsion–solid transfer (EST) process (Figure 5.7) was demonstrated for hydrogenations converting alkenes, alkynes, aromatic C=C bonds, cyano and nitro groups and recently extended to hydroformylation of alkenes with reaction yields from *ca.* 62 to 99%.[18]

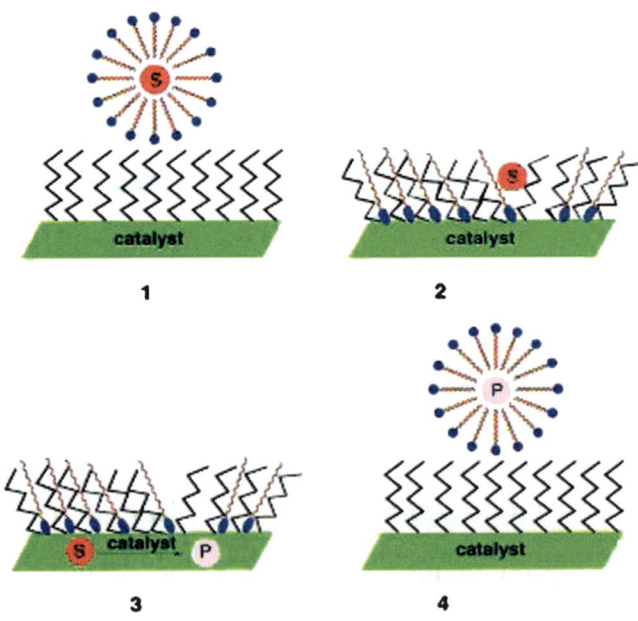

Figure 5.7   Transport, reaction and adsorption/desorption steps of the EST process. The emulsion which contains the substrate (1) spills its content into the catalyst material (2), the catalytic process takes place (3) and then the adsorbed surfactant carries the product back into solution (4). (Reproduced from ref. 18, with permission.)

Upon reaction, the heterogenized catalyst can be easily separated from the reaction mixture by filtration and then recycled. The hydrophobic substrate is microemulsified in water and subjected to an organometallic catalyst, which is entrapped within a partially hydrophobized sol–gel matrix. The surfactant molecules, which carry the hydrophobic substrate, adsorb/desorb reversibly on the surface of the sol–gel matrix breaking the micellar structure, spilling their substrate load into the porous medium that contains the catalyst. A catalytic reaction then takes place within the ceramic material to form the desired products that are extracted by the desorbing surfactant, carrying the emulsified product back into the solution.

A partially (10–20%) hydrophobized sol–gel yields optimal reactivity due to the fact that the organic groups tend to concentrate at the surface of the sol–gel cages imparting a high degree of lipophilic character to the material even at low concentration. A hydrophilic unmodified SiO_2 sol–gel matrix doped with the same catalyst indeed restricts reactants from being spilled *into* the sol–gel matrix (yielding abundant isomerization product); a more hydrophobic substrate readily adsorbs the reactants but does not *desorb* the products into the emulsion.

The key, thus, is careful control of the HLB of the solid itself along with *structural similarity* between the component of the catalyst and the substrate which seems to be necessary for successful catalytic reactions in the EST system. For example, hydrogenation of an arene was found to proceed best when the sol–gel matrix contained a phenyl ring, while an octyl-modified matrix was more suitable for the reduction of long-chained 1-octene.[19] The HLB, however, is not the only factor affecting reactivity of sol–gel catalysts, in that the *accessibility* and the *flexibility* of the cages entrapping the dopants are also relevant. For example, the versatile aerobic oxidation catalyst TPAP, which shows a modest activity in promoting alcohol oxidation when entrapped in a pure SiO_2 matrix, entrapped in partially hydrophobized silica xerogels becomes *more* active than the catalyst dissolved in toluene[20] (Figure 5.8).

Along with hydrophobicity, large amounts of both water (to promote hydrolysis) and methanol employed as co-solvent in the catalyst preparation (to promote homogeneity) are needed to ensure optimal reactivity, showing the number of experimental parameters of the sol–gel synthesis which can be controlled independently to optimize the performance of the resulting catalyst. Finally, in contrast to zeolites and other crystalline porous materials, amorphous sol–gel materials show a *distribution* of porosity which does not restrict the scope of application of sol–gel catalysts to substrates under a threshold molecular size.

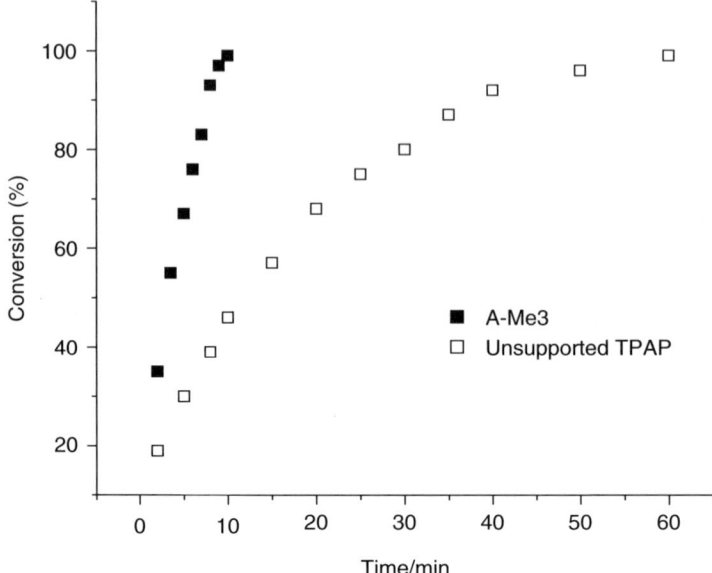

Figure 5.8   Oxidation kinetics in the aerobic conversion of benzyl alcohol to benzaldehyde in toluene mediated by 10 mol% TPAP either encapsulated in the
sol–gel hydrophobic matrix A-Me3 or unsupported. (Reproduced from
ref. 17, with permission.)

Hence, a single TPAP-doped ORMOSIL can be efficiently employed for
the oxidative dehydrogenation of very different alcohol substrates.

5.4   Chemically Entrapped Catalysts

Covalent bond immobilization of the catalyst to the sol–gel matrix is
usually sought to prevent leaching of the entrapped catalyst into the
reaction medium. As mentioned above, binding of the catalyst is easily
accomplished by using a silicon alkoxide derivatized with a suitable
ligand or with a functional group reacting with the dopant species
(Figure 5.9).

Site isolation of the entrapped catalyst offers unexpected properties
for supported homogeneous catalysts in a number of different conversions. Hence, for instance, in the synthesis of *N,N*-diethylformamide
from CO_2, H_2 and diethylamine, a mesoporous silica hybrid aerogel
doped with bidentate $RuCl_2[Ph_2P(CH_2)_3PPh_2]_2$ complexes exhibits
excellent selectivities (100%) and turnover frequencies up to $18\,400\,h^{-1}$,
more than six times *higher* than the corresponding homogeneous

Figure 5.9 TEMPO@DE is obtained by electrodeposition of a thin layer of organosilica doped with TEMPO (2,2,6,6-tetramethylpiperidine-1-oxyl) upon application of −1.1 V (*vs.* Ag/AgCl) for 15 min to a solution of suitable organosilanes (left). The electrocatalytic film thereby obtained selectively converts benzyl alcohol dissolved in 0.2 M NaHCO₃ (right).

catalyst.[20] Furthermore, the mesoporous areogel catalyst is almost nine times more active compared to a similarly doped microporous xerogel due to the high porosity of the former material, which enables easy access of the substrate to the active sites.

In contrast, a similarly doped Co(salen) aerogel (Figure 5.10) was *slow* in catalysing the oxidation of ethylbenzene to acetophenone despite showing quantitative conversion of ethylbenzene, a yield not possible with a similar heterogenized system obtained by conventional impregnation when a drastic reduction in activity is observed after 50% conversion.[21]

In other words, ligand silylation, affecting the *electronic* properties of the complex, may indeed result in a marked change of the activity of the entrapped complex. Silylation therefore should be planned carefully to synthesize the most suitable structure. This is shown in the example in Figure 5.11, where the hybrid organic–inorganic material results from the covalent attachment of manganese(III) porphyrin to a silica aerogel network. The macrocycle features three electron-withdrawing *meso*-2,6-dichlorophenyl groups, which stabilize metalloporphyrins against oxidative decomposition through steric and electronic effects, while the aminopropyltriethoxysilyl-functionalized   linker   ensures   optimal

OH + H_2N $Si(OEt)_3$ + $M(O_2CCH_3)_2.nH_2O$
EtOH
TEOS
EtOH
(i) HCl/ H_2O
(ii) N(hexyl)$_3$
$(EtO)_3Si$
$Si(OEt)_3$
links to the silica network

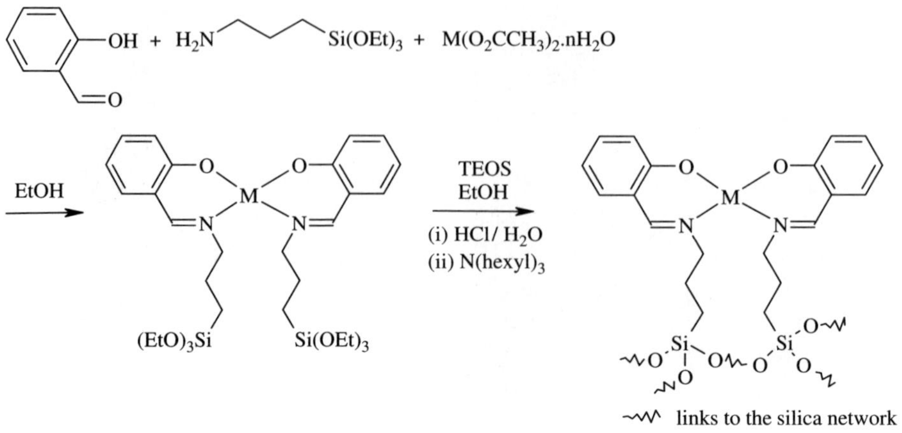

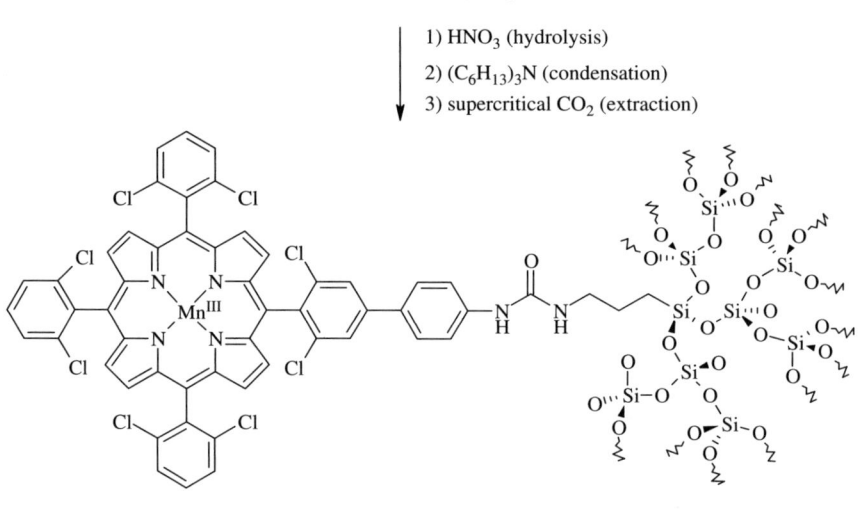

Figure 5.10 Sol–gel chemical encapsulation of a metal complex requires previous modification of the ligand. (Reproduced from ref. 21, with permission.)

Porphyrin + $Si(OMe)_4$ + H_2O
1) HNO_3 (hydrolysis)
2) $(C_6H_{13})_3N$ (condensation)
3) supercritical CO_2 (extraction)
Mn^{III}
links to the silica network

Figure 5.11 Entrapped in a silica aerogel, this manganese porphyrin is nine times more active than in solution. (Reproduced from ref. 22, with permission.)

covalent attachment to the final aerogel. As a result, the hybrid catalyst is highly active in the epoxidation of various olefins and in the hydroxylation of an alkane while, again, its activity is enhanced in comparison to the same porphyrin in homogeneous solution.[22]

As mentioned above with regard to entrapped TPAP, the flexibility of the encaging silica sol–gel cavities has a crucial role in ensuring optimal reactivity. Several examples show the general validity of this concept. Thus, nine phosphine–ruthenium complexes were chemically entrapped in silica and in different ORMOSIL and invariably the activity of the resulting catalysts in the hydrogenation of *n*-butanal increased with increasing mobility of the entrapped ruthenium complex,[23] *i.e.* with the degree of alkylation of the ORMOSIL. Notably, the organosilane $(CH_3)_2Si(OCH_3)_2$ was used as co-condensation precursor which is particularly effective in releasing the constraint imposed on the entrapped dopant by the silica sol–gel cage.

Leach-proof sol–gel entrapment can be exploited to carry out "one-pot" reactions with *mutually destructive* reactants while still allowing these reagents to activate or participate in desired reactions. For instance, three different one-pot redox reactions can be carried out in sequence in one pot over two separate sol–gel matrices doped with an oxidant (pyridinium dichromate) and with a reducing species $(RhCl[P(C_6H_5)_3]_3)$ without their mutual destruction and with no need for separation steps (Figure 5.12).[24]

The concept is general and one-pot acid/base and enzyme/catalyst enantioselective solid-state syntheses are easily achieved by entrapment of the mutually destructive reagents in two different sol–gel silicas. It is worth pointing out that while acids and bases adsorbed at the surface of polymers are left partly exposed and consequently require acid/base solid-state

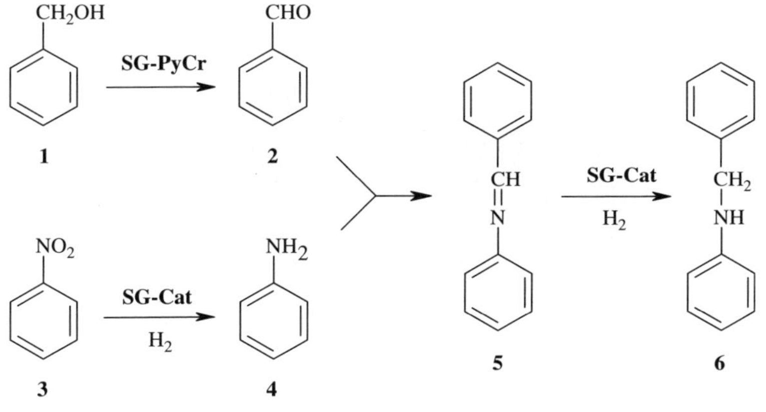

Figure 5.12   In this one-pot multistep synthesis, benzyl alcohol is first oxidized to benzaldehyde in a hydrogen-purged autoclave at 1 bar H_2. On raising the H_2 pressure to 13 bar, nitrobenzene is reduced to aniline which rapidly reacts with the aldehyde to form the Schiff base 5 in 91% yield. (Reproduced from ref. 24, with permission.)

syntheses to be carried out consecutively, their sol–gel entrapment allows smooth reaction in one pot and full recovery of the catalyst by simple filtration to be recycled in further runs without loss in catalytic activity.

In other cases, organic modification of the sol–gel cages markedly protects the entrapped molecular dopant from degradation by external reactants, as shown for instance by the entrapment of the radical 2,2,6,6-tetramethylpiperidine-1-oxyl (TEMPO). This is a highly active catalyst which in the NaOCl oxidation of alcohols to carbonyls in a CH_2Cl_2–H_2O biphasic system becomes highly stabilized upon sol–gel entrapment in an ORMOSIL matrix; it progressively loses it activity when entrapped at the external surface of commercial silica.[25]

Encapsulation starting from the readily available triacetonamine derivative 4-oxo-TEMPO and propylaminetrimethoxysilane, in fact, prevents the intermolecular quenching of the radicals bound at the silica surface that has been found to be responsible for the loss of activity of TEMPO tethered at the surface of commercial silica.

Once again encapsulation of the dopant does the trick by restricting the modes of access of incoming reactants to the catalyst entrapped at the cage surfaces. Thus, while homogeneous TEMPO *fails* to oxidize benzylic aminoalcohols to aminohydroxyacids, a fully methylated sol–gel ORMOSIL doped with TEMPO selectively affords these valuable pharmaceutical precursors in high yields (60–80%; Figure 5.13) by isolating the nitroxyl radicals within the hydrophobized pores (Figure 5.14).[26]

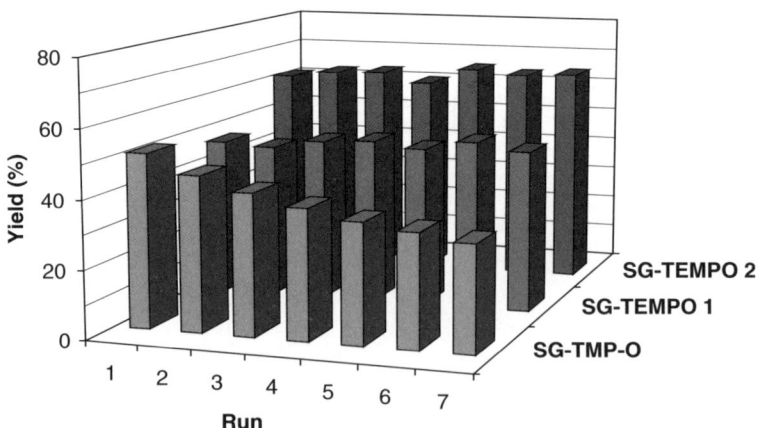

Figure 5.13   Yields in the Montanari–Anelli oxidation of 1-nonanol to give nonanal in the presence of silica-supported TEMPO (SG-TMP-O, front row), and in the presence of sol–gel ORMOSIL doped with TEMPO [SG-TEMPO-1 is 25% and SG-TEMPO-2 is 100% methylated (middle and back row, respectively)]. (Reproduced from ref. 25, with permission.)

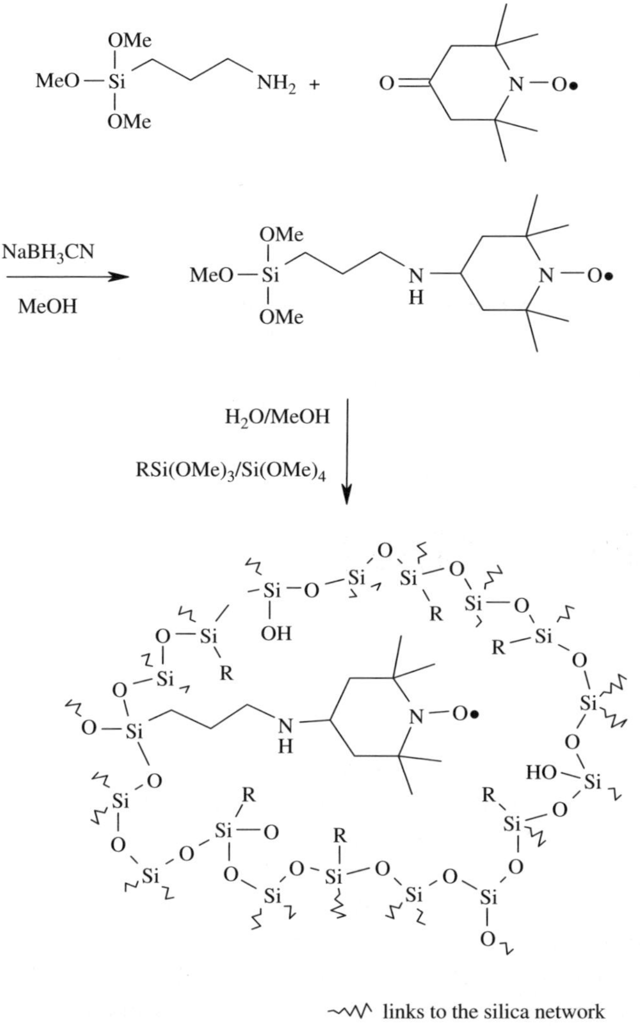

~wwv links to the silica network

Figure 5.14 Sol–gel immobilized TEMPO is an "off-the-shelf" alcohol oxidation catalyst. In a biphasic reaction system and in organic solvent it yields carbonyls; in water it yields carboxylates.

5.5 Biocatalysis

Biological species such as enzymes, whole cells, antibodies and even bacteria can all be successfully entrapped in silica sol–gel matrices, often with enhancement of activity with respect to the free biologicals. In these cases, the process is adapted to eliminate toxic alcohols which are typically released in conventional sol–gel processes based on the hydrolysis of silicon alkoxides. Two such methods are the use of silicon alkoxide

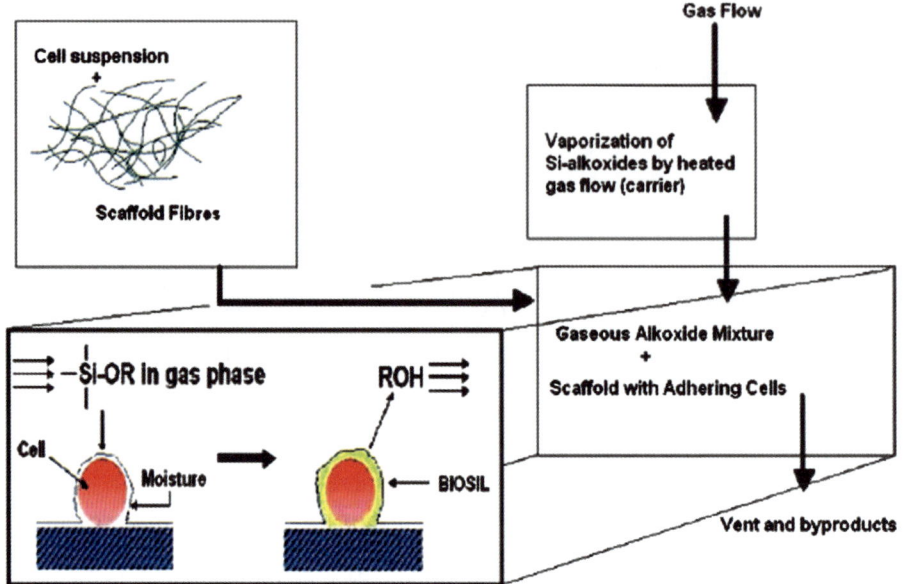

Figure 5.15   The Biosil process. Si–OR: alkoxide precursors; ROH: reaction
by-products. (Reproduced from ref. 27b, with permission.)

precursors with glycerol-derived polyol silicates and the "Biosil" process
of spraying aqueous tetraethylorthosilicate over a suspension of living
cells to cover them with a thin siliceous layer (Figure 5.15).[27]

The conceptual basis of the approach—the use of functional cells or
cell aggregates as a highly specialized laboratory for producing the
desired substances—has enormous practical consequences: cells indeed
are a complete and natural system encompassing the entire enzymatic
chain and yielding specific molecules from elemental substrates. Furthermore, the process is general and versatile: gas-phase silicon alkoxides
react with the wet surface of cells, affording a mechanically stable and
homogeneous layer of amorphous SiO_2 modified by Si–R and Si–H
bonds. The layer does not suppress cell viability or functionality, while,
allowing porosity control, it provides important immunological protection (by tailored exclusion of access to macromolecules above a certain
threshold pore size). A number of pharmaceutical molecules, including
valued paclitaxel (Taxol), are currently being produced at a small scale
over sol–gel protected cells. Italy-based company IRB (Istituto di
Ricerche Biotecnologiche) aims to license the technology and start production of Taxol and other anticancer drugs over sol–gel entrapped cells
of *Taxus brevifolia*. Similarly, thiol-functionalized organosilica microspheres synthesized *via* a novel surfactant-free emulsion-based method

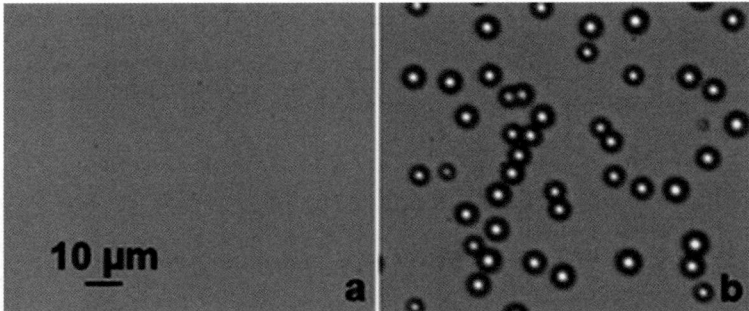

Figure 5.16 Formation of emulsion droplets. (a) Aqueous MPS solution after acid-catalysed hydrolysis and condensation. (b) Micrometre-sized emulsion droplets are rapidly formed upon addition of the base catalyst triethanolamine. (Reproduced from ref. 28, with permission.)

show great potential for optical encoding and in biomolecular screening applications.[28]

The microspheres—synthesised *via* a two-step process (acid-catalysed hydrolysis and condensation of 3-mercaptopropyltrimethoxysilane (MPS) in aqueous solution, followed by condensation catalysed by triethanolamine)—have a narrow size distribution (Figure 5.16) and are considerably more stable than polystyrene–divinylbenzene microspheres as shown in phosphoramidite oligonucleotide synthesis by the excellent retention of fluorescence intensity in each of the reagent steps involved in phosphoramidite DNA synthesis (Figure 5.17, in which the organo-silica microsphere free thiol groups are derivatized with ATTO 550 maleimide coupled to the entrapped dye).

A similar major advancement that will greatly speed up the drug discovery process is due to efficient biocatalysis over ORMOSIL-entrapped cytochrome enzyme P450.[29] By simply spotting a precursor solution of P450 in MTMS and aqueous HCl over a MTMS-coated glass slide, a device is obtained (MetaChip) that combines high-throughput P450 catalysis with cell-based screening on a microscale platform (Figure 5.18).

This technology, demonstrated by using sol–gel encapsulated P450 to activate the prodrug cyclophosphamide (a major constituent of the anticancer drug Cytoxan), allows rapid and inexpensive toxicity assessment at the early phases of drug development, providing a high-throughput microscale alternative to currently used *in vitro* methods for human metabolism and toxicology screening based on liver slices, cultured human hepatocytes or isolated P450 itself. Here, again, the fully alkylated silica matrix provides the cage flexibility necessary for the

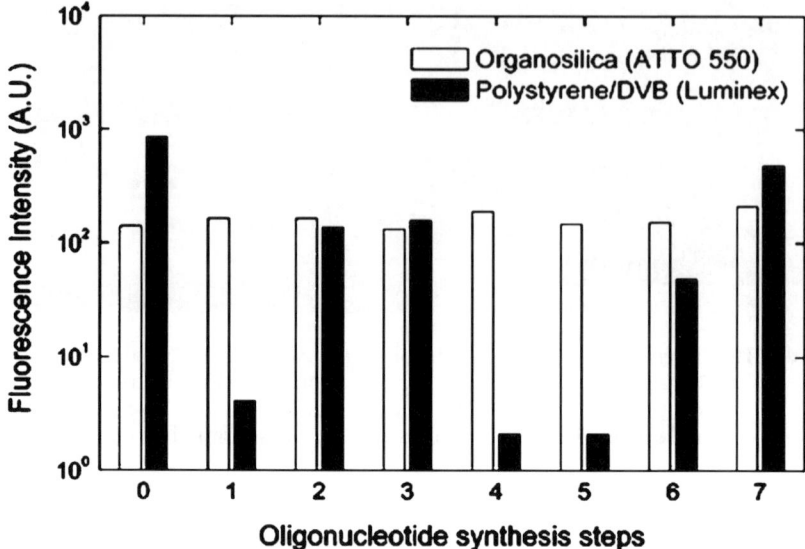

Figure 5.17 Stability of optical encoding towards oligonucleotide reagents. Organosilica microspheres covalently labelled with ATTO 550 dye are stable
towards each of the reagents used in phosphoramidite oligonucleotide
synthesis. In contrast, optically encoded polystyrene–divinylbenzene
(DVB) beads are unstable in most steps, in particular those involving
dichloromethane and tetrahydrofuran. (Reproduced from ref. 28, with
permission.)

enzyme to change conformation and react with the incoming reactants,
which, in this case, are the cell metabolites; 45% of the enzyme activity
in solution is retained upon encapsulation in the sol–gel ORMOSIL
film, regardless of the fact that the enzyme is entrapped starting from an
acidic solution and that toxic methanol is released in the sol–gel polycondensation of the monomers.

5.6   Commercial Catalysts and Forthcoming Applications

Catalytic sol–gel lipase immobilizates were rapidly commercialized (by
Fluka) after their invention in 1995 because of their remarkably stable
activity in esterification reactions (and also in the kinetic resolution of
chiral alcohols and amines) along with unique stability (residual activity
of 70% even after 20 reaction cycles is common). The original procedure for the encapsulation produced by the fluoride-catalysed hydrolysis
of mixtures of $RSi(OCH_3)_3$ and $Si(OCH_3)_4$ has been improved

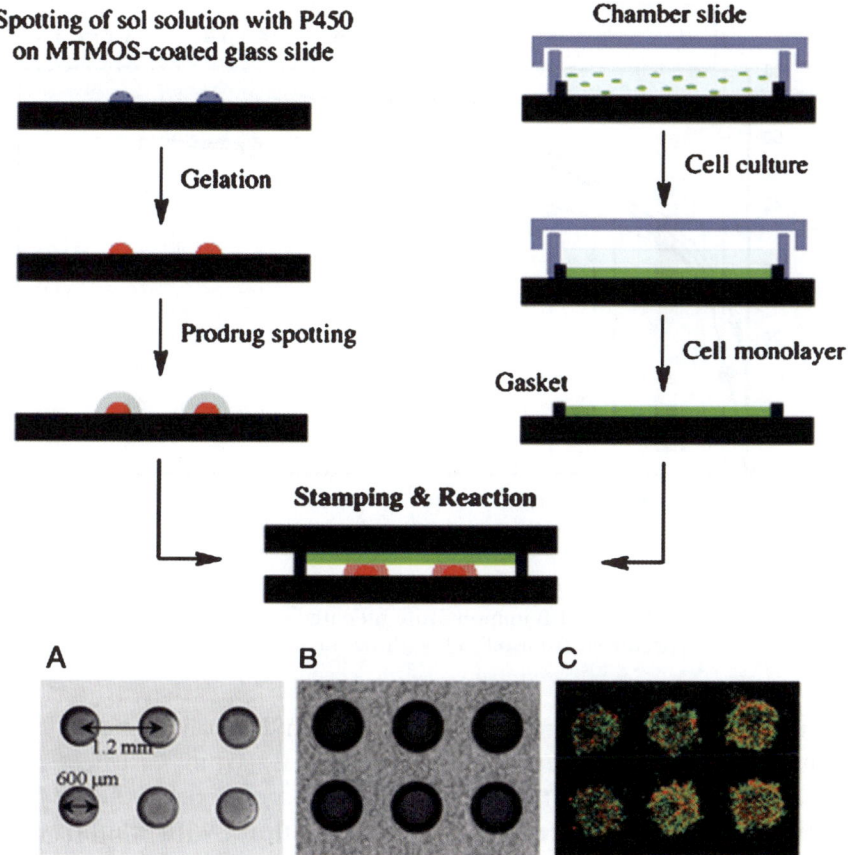

Figure 5.18 Schematic of MetaChip platform and microscopy images of sol–gel spots. Shown are (A) 30 nl P450 sol–gel spots, (B) 30 nl sol–gel spots with 60 nl of prodrug solution after being stamped by a cell monolayer and (C) cell monolayer after removal from sol–gel array and staining. (Reproduced from ref. 29, with permission.)

considerably with higher enzyme loading, variation of the alkylsilane precursor and the use of additives, and these materials have now reached a second-generation level of performance.[30] The materials are also excellent recyclable catalysts in the kinetic resolution of chiral alcohols and amines (residual activity of 70% even after 20 reaction cycles; Figure 5.19).

On a laboratory scale, for example, the best performing catalyst for esterification requires double immobilization of lipase within the cages of an ORMOSIL (i-C_4H_9–TMOS = 5:1) and further coated on the external surface of Celite along with co-entrapment of 18-crown-6 ether

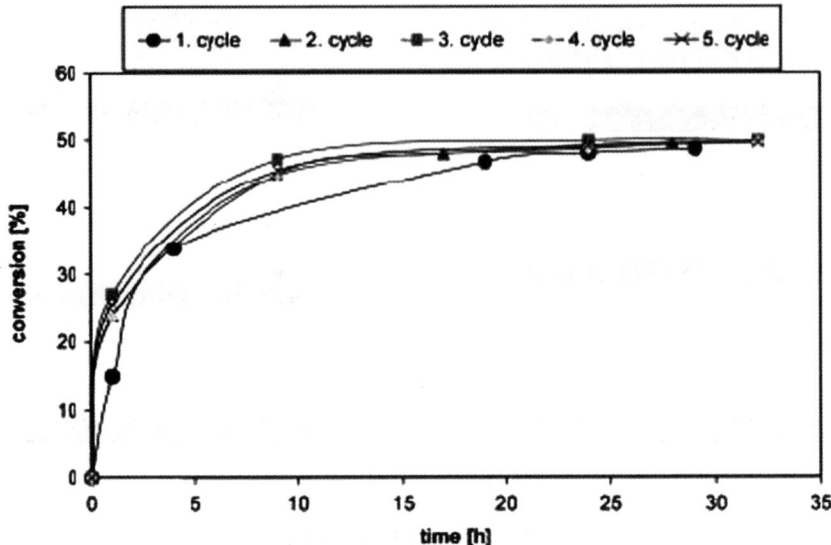

Figure 5.19   Recycling experiments in the kinetic resolution of racemic amine using
the sol–gel CaLB immobilizate prepared with 18-crown-6 as an additive.
(Reproduced from ref. 30, with permission.)

in the presence of small amounts of polyvinyl alcohol and isopropyl
alcohol. Using sol–gel encapsulated lipase B (from CaLB, *Candida
antarctica*) prepared in the presence of 18-crown-6 in the acylating
kinetic resolution of racemic 1-phenylethylamine with methoxyacetic
acid ethyl ester as acylating agent, the desired (*S*)-amine was isolated
enantiomerically pure (ee >99%), while 50% conversion was reached
within 29 h, *i.e.* the highly enantioselective reaction ideally stops when
the (*R*)-enantiomer has been consumed. Figure 5.19, showing the results
of five cycles, demonstrates a constant performance with respect to
activity with no reduction in enantioselectivity in any of the runs.

The high catalyst loading typical of sol–gel entrapped catalysts
ensures a desirably high substrate/catalyst (S/C) ratio as the major part
of the heterogeneous catalyst weight originates from the silicate matrix.
For example, in a preparative-scale reaction of the alcohol *rac*-1-(2-
naphthyl)-ethanol only 250 mg of sol–gel CaLB immobilizate could be
used per 10 g of substrate. For comparison, all this makes the process
based on sol–gel immobilized lipase very competitive with the commercial BASF process using lipase immobilized on Amberlite to produce
the amine at a scale of 1000 tons per year.

The fundamental sol–gel biocatalytic process has been licensed by
IRB to Italian company Indena. The aim is to start production of

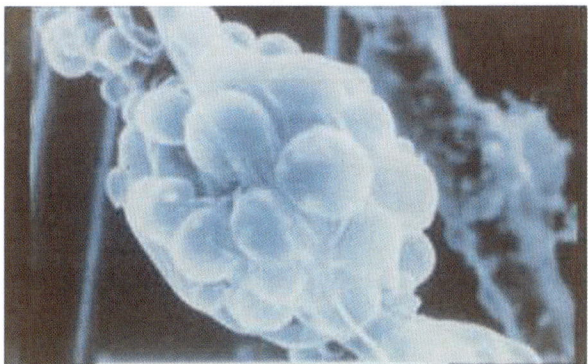

Figure 5.20   Sol–gel silica entrapped cells such as those of *Saccharomyces cerevisiae* shown here are highly stabilized and freely accessible to external nutrients acting as potent bioreactors. (Photo courtesy of Giovanni Carturan.)

paclitaxel over silica-entrapped *Taxus* cells in the bioreactor system mentioned above, affording a precious anticancer drug without having to destroy the *Taxus* trees from whose bark the compound is extracted in very low yields. The process is general in that it prevents cell death and affords a whole set of bioactive materials (Figure 5.20) whose enormous potential awaits full exploitation in catalysis and in many other fields. IRB indeed has achieved the generation of a large library of plant cell lines including *Syringa vulgaris*, *Ajuga reptans* and many others with the production of secondary metabolites useful in a number of promising commercial applications. A pilot production facility is currently being used for the production of plant cell biomass by employing innovative biological reactors, optimized to increase cell growth and productivity.

The British fine chemicals manufacturer Avecia licensed Johnson Matthey's sol–gel technology for preparing organic–inorganic silica hybrid gels doped with chiral ligands (CACHy, catalytic asymmetric cyanohydrin) to encapsulate catalytic salen complexes of vanadium or titanium. The latter catalysts[31] allow easy conversion of aldehydes and ketones into chiral cyanohydrins, high value building blocks and useful precursors for hydroxyamino acids and amino alcohols (Figure 5.21).

Chiral cyanohydrins in fact are high-value building blocks and useful precursors for hydroxyamino acids and amino alcohols. The new sol–gel immobilization technology licensed from Johnson Matthey and named CTIS CACHy (CTIS: Chiral Technologies Interface System)[32] dramatically improves process economics for large-scale pharmaceutical manufacturing as it increases the turnover number of the catalyst

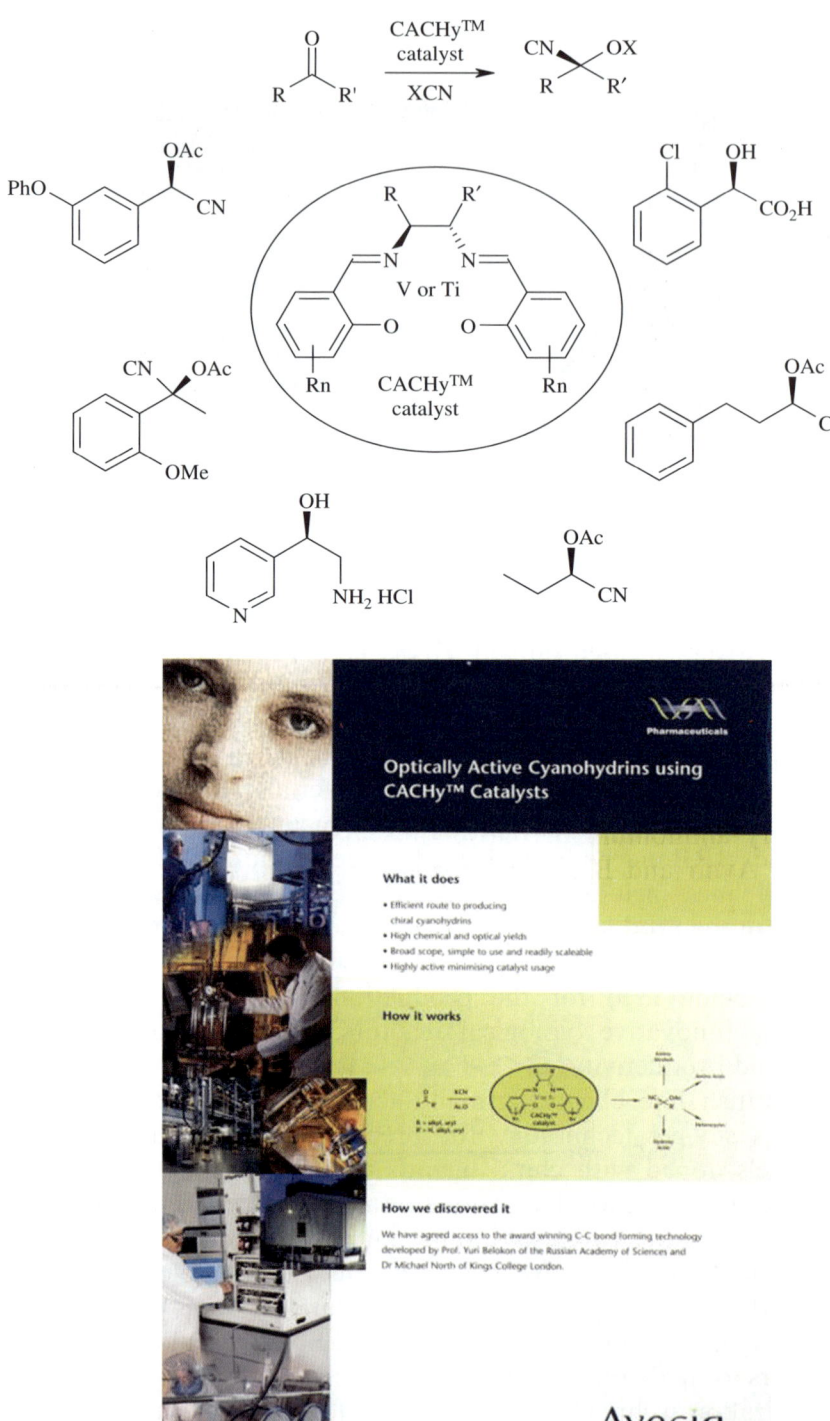

Figure 5.21 Cyanohydrins are precursors for hydroxyamino acids, amino acids and amino alcohols. New sol–gel CACHy catalysts convert aldehydes and ketones into such valued chiral building blocks. (Reproduced from Avecia.com, with permission.)

compared with its homogeneous version, while maintaining enantiomeric excess above 90%. Clearly, the competitive advantage offered here by these sol–gel catalysts, new and lower cost cyanide reagents, will push many other companies to adopt commercial sol–gel catalysts for their fine chemicals productions, while Johnson Matthey itself is continuing to develop a line of chiral sol–gel catalytic materials in a joint development with Avecia.

More recently, a series of sol–gel hydrophobized nanostructured silica matrices doped with the organocatalyst TEMPO (SiliaCat TEMPO) entered the market as suitable oxidation catalysts for the rapid and selective production of carbonyls and carboxylic acids. In the former case, SiliaCat TEMPO selectively mediates the oxidation of delicate primary and secondary alcohol substrates into valued carbonyl derivatives (Scheme 5.2), retaining its potent activity throughout several reaction cycles (Table 5.2).[33] Using this catalyst, for example, enables the synthesis of extremely valuable α-hydroxy acids with relevant selectivity enhancement by coupling of SiliaCat TEMPO with rapid RuO_4-mediated olefin dihydroxylation (Scheme 5.3).[34]

Several sol–gel entrapped catalysts are likely to soon find commercial applications. A variety of transition metal catalysts physically entrapped in silica matrices as ion pairs generated from the metal halides and quaternary ammonium or phosphonium salts developed in the mid-1990s by Avnir and Blum resulted in truly heterogeneous, stable and

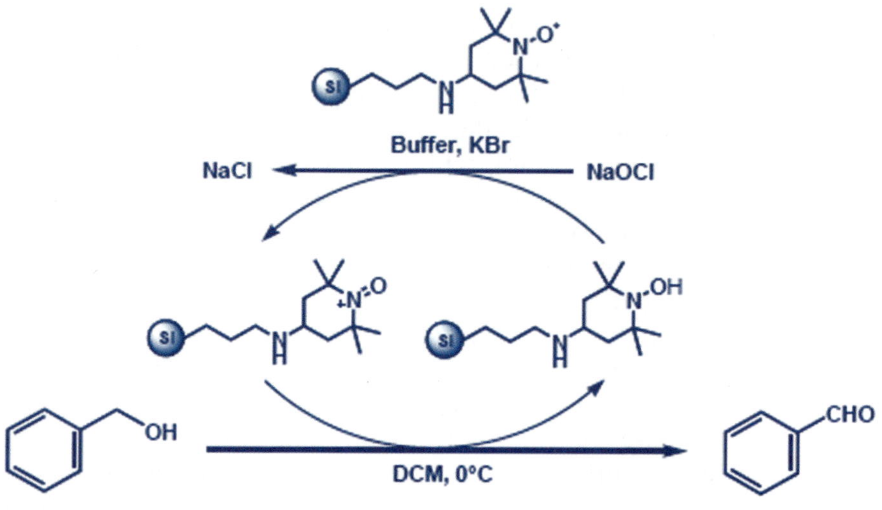

Scheme 5.2 SiliaCat TEMPO catalytic cycle.

Table 5.2 Activity of SiliaCat TEMPO in consecutive reaction cycles. (Reproduced from ref. 34, with permission.)

Recycle	Time (min)	Yield (%)
1st to 6th	30	100
7th	30/60	88/96
8th	30/60	95/100
9th	30/60	97/100
10th	30/60	90/100

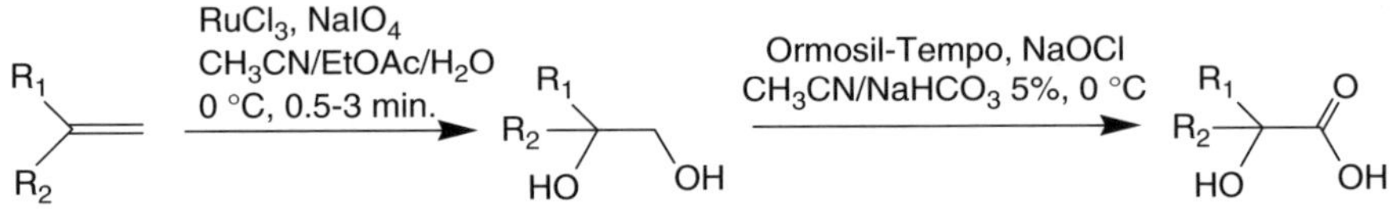

Scheme 5.3 Synthesis of α-hydroxy acids.

selective mediators for a number of reactions, including hydrogenation, hydroformylation and double bond migration conversions.[35]

The resulting sol–gel catalysts usually proved more stable and *versatile* under ambient conditions than their homogeneous analogues. For example, a remarkable asymmetric hydrogenation of prochiral itaconic acid over sol–gel entrapped (−)-Ru-BINAP in water becomes possible which simply cannot be done with non-entrapped, water-insoluble catalysts.

Another important highly selective and stable hydroformylation sol–gel catalyst is made of silica-supported rhodium covalently bound to supported Xantphos family of ligands.[36] By incorporating monoliths of the sol–gel doped material into the paddles of an autoclave stirrer, the catalyst (Rotacat) can be used in a *continuous* liquid flow process. A single sample of this catalyst was used for a variety of different hydroformylation reactions under widely varying conditions over a period of more than a year, still retaining its selective activity.

Along with the frequently higher activity, selectivity and stability of sol–gel molecular catalysts reported above, another major advantage which adds value to the performance of these materials is their broad applicability to novel reaction systems. Industrial conversions in supercritical carbon dioxide, for instance, are carried out in a relatively small, high-throughput continuous reactor in which the catalyst is packed at the bottom of the reactor and the solvent continuously recycled. In the case of this known fluorophilic solvent, the sol–gel process may easily afford hydrophobized fluorinated catalytic silica gels[37] which can be shaped in practically any form (film, coating, fibre, foam, powder, *etc.*) to meet the engineering requirements of a continuous process. They can then be applied to the aerobic oxidation of alcohols into valuable carbonyls eliminating any by-products while carrying out the reaction in a non-flammable, non-toxic and recyclable reaction medium.

References

1. This accounts for the importance of the concept of atom efficiency pioneered by Barry Trost and Roger Sheldon in the early 1990s. For a number of applicative cases, see: S. K. Ritter, *Chem. Eng. News*, 2002, **80**, 19.
2. B. W. Cue, *Chem. Eng. News*, 2005, **83**, 46.
3. S. J. Broadwater, S. L. Roth, K. E. Price, M. Kobaslija and D. T. McQuade, One-pot multi-step synthesis: a challenge spawning innovation, *Org. Biomol. Chem.*, 2005, **3**, 2899.
4. D. J. Cole-Hamilton, Homogeneous Catalysis–New Approaches to Catalyst Separation, Recovery, and Recycling, *Science*, 2003, **299**, 1702.
5. S. Kobayashi and R. Akiyama, Renaissance of immobilized catalysts. New types of polymer-supported catalysts, micro-encapsulated catalysts, which enable environmentally benign and powerful high-throughput organic synthesis, *Chem. Commun.*, 2003, 449.
6. Detailed information on the technology and its applications is available at: http://www.reaxa.co.uk.
7. This catalyst consists of polymer fibres functionalized *via* graft copolymerization to which metal species are covalently bound resulting in a high density of active and accessible functional sites enabling efficient heterogenization of a number of homogeneous catalysts with high loading ($\sim$2–3 mmol g^{-1}). The fibres are commercialized by Johnson Matthey in a variety of lengths (0.25 mm is standard) giving materials with different physical characteristics suited to a variety of reactor and filter combinations. For example: M. Gilhespy, M. Lok and X. Baucherel, Polymer-supported nitroxyl radical catalyst for selective aerobic oxidation of primary alcohols to aldehydes, *Chem. Commun.*, 2005, 1085.
8. (a) R. Ciriminna and M. Pagliaro, Recent Uses of Sol–Gel Doped Catalysts in the Fine Chemicals and Pharmaceutical Industry, *Org. Process Res. Dev.*, 2006, **10**, 320; (b) R. Ciriminna and M. Pagliaro, Catalysis by Sol-Gels: An Advanced Technology for Organic Chemistry, *Curr. Org. Chem.*, 2004, **8**, 1851.

9. R. Ciriminna, L. M. Ilharco, A. Fidalgo, S. Campestrini and M. Pagliaro, The structural origins of superior performance in sol–gel catalysts, *Soft Matter*, 2005, **1**, 231.

10. As shown, for instance, by the lack of any sol–gel catalyst cited in a recent, thorough review covering tens of industrial applications of sol–gel hybrid materials: C. Sanchez, B. Julián, P. Belleville and M. Popall, Applications of hybrid organic–inorganic nanocomposites, *J. Mater. Chem.*, 2005, **15**, 3559.

11. M. Pagliaro, R. Ciriminna, M. Wong Chi Man and S. Campestrini, Better Chemistry through Ceramics: The Physical Bases of the Outstanding Chemistry of ORMOSIL, *J. Phys. Chem. B*, 2006, **110**, 1976.

12. J. Blum and D. Avnir, Catalysis and reactivity with sol-gel entrapped organic and organometallic chemicals, in *Handbook of Sol-Gel Science and Technology*, ed. S. Sakka, Springer, New York, 2003, vol. III, ch. 24.

13. S. J. Broadwater and D. T. McQuade, Investigating PdEnCat Catalysis, *J. Org. Chem.*, 2006, **71**, 2131.

14. M. D. Jones, R. Raja, J. M. Thomas, B. F. G. Johnson, D. W. Lewis, J. Rouzaud and K. D. M. Harris, Enhancing the Enantioselectivity of Novel Homogeneous Organometallic Hydrogenation Catalysts, *Angew. Chem. Int. Ed.*, 2003, **42**, 4326.

15. D. Avnir, Organic Chemistry within Ceramic Matrixes: Doped Sol–Gel Materials, *Acc. Chem. Res.*, 1995, **28**, 328.

16. A. Fidalgo and L. M. Ilharco, Chemical Tailoring of Porous Silica Xerogels: Local Structure by Vibrational Spectroscopy, *Chem. A. Eur. J.*, 2004, **10**, 392.

17. R. Ciriminna and M. Pagliaro, Tailoring the Catalytic Performance of Sol-Gel-Encapsulated Tetra-n-propylammonium Perruthenate (TPAP) in Aerobic Oxidation of Alcohols, *Chem. Eur. J.*, 2003, **9**, 5067.

18. R. Abu-Reziq, J. Blum and D. Avnir, Three-Phase Microemulsion/Sol-Gel System for Aqueous Catalysis with Hydrophobic Chemicals, *Chem. Eur. J.*, 2004, **10**, 958.

19. (a) R. Abu-Reziq, J. Blum and D. Avnir, Three-Phase Microemulsion/Sol–Gel System for Aqueous Catalytic Hydroformylation of Hydrophobic Alkenes, *Eur. J. Org. Chem.*, 2005, **17**, 3640; (b) R. Abu-Reziq, D. Avnir and J. Blum, A Three-Phase Emulsion/Solid-Heterogenization Method for Transport and Catalysis, *Angew. Chem. Int. Ed.*, 2002, **41**, 4132.

20. L. Schmid, M. Rohr and A. Baiker, A mesoporous ruthenium silica hybrid aerogel with outstanding catalytic properties in the synthesis

of N,N-diethylformamide from CO_2, H_2 and diethylamine, *Chem. Commun.*, 1999, **21**, 2303.

21. E. F. Murphy, L. Schmid, T. Burgi, M. Maciejewski, A. Baiker, D. Gunther and M. Schneider, Nondestructive Sol–Gel Immobilization of Metal (salen) Catalysts in Silica Aerogels and Xerogels, *Chem. Mater.*, 2001, **13**, 1296.

22. M. Bonnet, L. Schmid, A. Baiker and F. Diederich, A New Mesoporous Hybrid Material: Porphyrin-Doped Aerogel as a Catalyst for the Epoxidation of Olefins, *Adv. Funct. Mater.*, 2002, **12**, 39.

23. E. Lindner, A. Jaeger, F. Auer, W. Wielandt and P. Wegner, Supported organometallic complexes. Part XIII. Catalytic studies on sol–gel processed (ether phosphine)ruthenium(II) complexes with different spacer lengths and different polysiloxane matrices, *J. Mol. Catal. A: Chem.*, 1998, **129**, 91.

24. F. Gelman, J. Blum and D. Avnir, One-pot sequences of reactions with sol-gel entrapped opposing reagents. Oxidations and catalytic reductions, *New J. Chem.*, 2003, **27**, 205.

25. R. Ciriminna, C. Bolm, T. Fey and M. Pagliaro, Sol–Gel Ormosils Doped with TEMPO as Recyclable Catalysts for the Selective Oxidation of Alcohols, *Adv. Synth. Catal.*, 2002, **344**, 159.

26. M. L. Testa, R. Ciriminna, C. Hajji, E. Zaballos Garcia, M. Ciclosi, J. Sepulveda Arques and M. Pagliaro, Oxidation of Amino Diols Mediated by Homogeneous and Heterogeneous TEMPO, *Adv. Synth. Catal.*, 2004, **346**, 655.

27. (a) E. M. Cappelletti, G. Carturan and A. Piovan, Production of secondary metabolites with plant cells immobilized in a porous inorganic support, *US Pat.*, 5 998 162, 1999. (b) See also the review: G. Carturan, R. Dal Toso, S. Boninsegna and R. Dal Monte, Encapsulation of functional cells by sol–gel silica: actual progress and perspectives for cell therapy, *J. Mater. Chem.*, 2004, **14**, 2087.

28. C. R. Miller, R. Vogel, P. P. T. Surawski, S. R. Corrie, A. Rühmann and M. Trau, Biomolecular screening with novel organosilica microspheres, *Chem. Commun.*, 2005, 4783.

29. M.-Y. Lee, C. B. Park, J. S. Dordick and D. S. Clark, Metabolizing enzyme toxicology assay chip (MetaChip) for high-throughput microscale toxicity analyses, *Proc. Natl. Acad. Sci.*, 2005, **102**, 983.

30. M. T. Reetz, P. Tielmann, W. Wiesenhofer, W. Konen and A. Zonta, Second Generation Sol–Gel Encapsulated Lipases: Robust Heterogeneous Biocatalysts, *Adv. Synth. Catal.*, 2003, **345**, 717.

31. A. J. Blacker, M. North and Y. N. Belokon, *Chem. Today*, 2004, **22**, 30.
32. M. Rouhi, *Chem. Eng. News*, 2004, 82, 20. See also the Avecia press release of 16 January 2004 at: www.avecia.com.
33. A. Michaud, G. Gingras, M. Morin, F. Béland, R. Ciriminna, D. Avnir and M. Pagliaro, SiliaCat TEMPO: An Effective and Useful Oxidizing Catalyst, *Org. Process Res. Dev.*, 2007, **11**, 766.
34. P. Gancitano, R. Ciriminna, M. Luisa Testa, A. Fidalgo, L. M. Ilharco and M. Pagliaro, Enhancing selectivity in oxidation catalysis with sol–gel nanocomposites, *Org. Biomol. Chem.*, 2005, **3**, 2389.
35. J. Blum, D. Avnir and H. Schumann, *Chemtech*, 1999, **29**, 32. In a typical series of experiments at 80 °C and 10 atm H_2 using 2.2 mmol substrate and 3.1×10^{-2} mmol sol–gel entrapped (−)-Ru-BINAP in 4 ml H_2O, the yields (and o.p.) of the resulting (+)-2-methylsuccinic acid in the first four runs of 24 h were 100 (52), 98 (50), 95 (46) and 90% (41%), respectively.
36. A. J. Sandee, J. N. H. Reek, P. C. J. Kamer and P. W. N. M. van Leeuwen, A Silica-Supported, Switchable, and Recyclable Hydroformylation–Hydrogenation Catalyst, *J. Am. Chem. Soc.*, 2001, **123**, 8468.
37. R. Ciriminna, S. Campestrini and M. Pagliaro, Fluorinated Silica Gels Doped with TPAP as Effective Aerobic Oxidation Catalysts in Dense Phase Carbon Dioxide, *Adv. Synth. Catal.*, 2004, **346**, 231.

CHAPTER 6
Sensing

6.1 Optical Sol–Gel Sensors

Chemical sensing with doped sol–gel glasses is, simply, the most versatile and potent sensing technology available today and sol–gel sensors are rapidly replacing older detection devices in industry, medicine, environmental monitoring and in general in all the contexts where rapid and accurate measurement of one or more analytes is required (Figure 6.1).

The common working principle of sol–gel sensors is simple and has been applied to a large number of new sensors since the development of the doped sol–gel methodology for chemical sensing purposes in the late 1980s and early 1990s.[1] The advantages over a traditional commercial electrochemical sensor are well illustrated by the comparison between a dye-doped xerogel oxygen sensor and the older electrodes reported in Table 6.1.

Chemical changes in the environment of the entrapped molecule are sensed and revealed through colour-developing reactions between the entrapped reagent and an external diffusable chemical species *or* through the emission of light. Reactions include traditional chemical proton transfer reactions (such as pH indicators), redox reactions, complexations and ligand exchanges as well as biochemical enzymatic reactions to reveal analytes such as H^+, metal ions, inorganic anions, glucose, oxygen and ammonia.

The chemical inertness and optical transparency of SiO_2 sol–gel matrices make the doping methodology ideal for sensor developments, even if leaching is an issue that must be often addressed. Another attractive, unique feature is that non-toxic sol–gel silicas can be made in

Silica-Based Materials for Advanced Chemical Applications
By Mario Pagliaro
© Mario Pagliaro 2009
Published by the Royal Society of Chemistry, www.rsc.org

Figure 6.1   FOXY oxygen sensor probe. (Photo courtesy of Ocean Optics, Inc.)

various forms as required by the analytical method and, in particular, they can be miniaturized and integrated (coated) into fibre optics affording sensors ideally suited for remote sensing and *in vivo* measurement in harsh environments, in the food industry and in medical diagnoses (Figure 6.2).[2]

Being a high surface area adsorbent, a sol–gel silica matrix concentrates by adsorption an impurity and then detects it, greatly enhancing the sensitivity of conventional solution chemistries. This was first demonstrated in 1990 in the case of Fe^{2+} sensing by sol–gel SiO_2-entrapped *o*-phenanthroline.[3] Sensitivity then is about two orders of magnitude better than conventional solution absorption spectroscopy, reaching an impressive detection limit of 100 ppt. Since sol–gel glass is compatible with various sensing reagents, multiple sensors (*i.e.* sensor array) and sensors containing several sensing reagents can be developed.

Methods of indicator immobilization in sol–gels include physical and chemical (covalent binding) doping by incorporation of an indicator or reagent during formation of the sol–gel glass.

Table 6.1 Comparison between fibre-optic oxygen sensors and commercial electrode sensors.

Fibre-optic oxygen sensor systems	*Commercial electrodes*
Based on dynamic equilibrium with surrounding media; O_2 is not consumed	Measure the rate of consumption of O_2 which is indirectly related to concentration in surrounding media; also respond to stirring speed, viscosity and condition of the membrane
Respond to pO_2; calibration is the same for both gases and liquids	Can be calibrated for use in gases or liquids, but not both at the same time
Immune to sample matrix, and changes in pH, salinity and ionic strength of environment	Polarographic electrodes are affected by sample matrix and must be calibrated using standards that are similar to the samples; changes in pH, salinity and ionic strength of environment affect their readings
Immune to interference from moisture, carbon dioxide, methane and other substances	Electrochemical electrodes are subject to interference from a number of substances and sampling conditions
Fast response time: <1 s for dissolved oxygen and oxygen gas	Can have a response time of 60–90 s, depending on temperature, sample viscosity and stirring speed
Long life: more than a year	Typical lifetime of just 3 months; membranes must be changed frequently as they become fouled, damaged or clogged
Frequent calibration is unnecessary	Calibration may be necessary on an hourly basis, or any time the viscosity or stirring speed of the sample changes
Probe temperature range is −80 to +80 °C (up to 110 °C for brief periods only)	Temperature range for some electrodes is only 0–45 °C

In general, comparison between physical and chemical doping by covalent bonding shows that whereas the long-term stability of sensors based on covalently bound indicators is unsurpassed, they often display relatively small signal changes and rather long response times. On the other hand, leaching is the main problem in physically doped glasses. This is clearly shown by the pH indicator dye aminofluorescein (AF) doped into a SiO_2 sol–gel which is an adequate material for sensing pH in terms of response time, relative signal change and stability, even though leaching is a problem if the sensor is to be operated continuously.[4]

Plots in Figure 6.3 show leaching and response of pH sensor layers chemically doped starting from 3-(trimethoxysilyl)propylisocyanate (ICPS) and (glycidyloxypropyl)trimethoxysilane (GOPS) derivatized with the pH indicator AF, or physically doped in SiO_2 (from tetramethylorthosilicate, TMOS) and in 50% phenyl-modified silica.

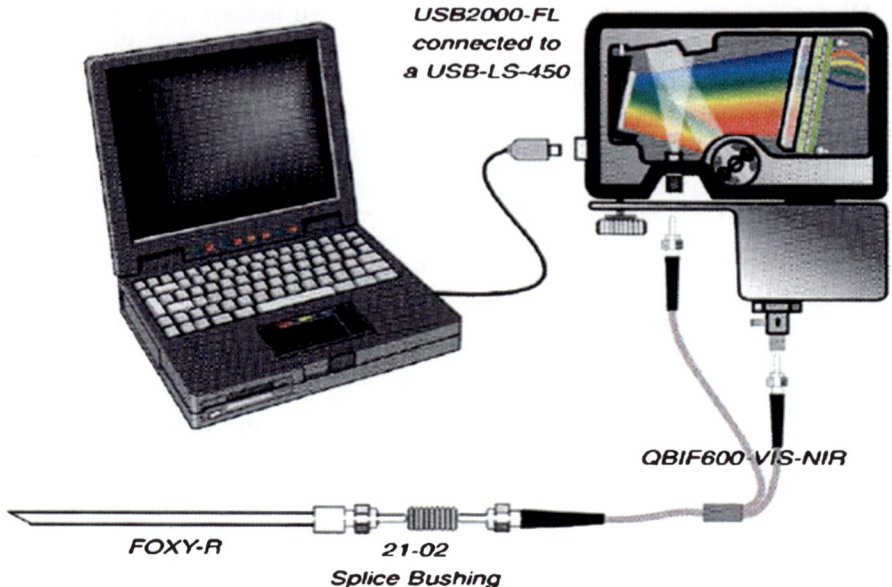

Figure 6.2 A bifurcated optical fibre is used to carry light to a probe. The hose tip consists of a thin layer of hydrophobic sol–gel material doped with a ruthenium complex which senses oxygen and reveals its content. (Image courtesy of Ocean Optics.)

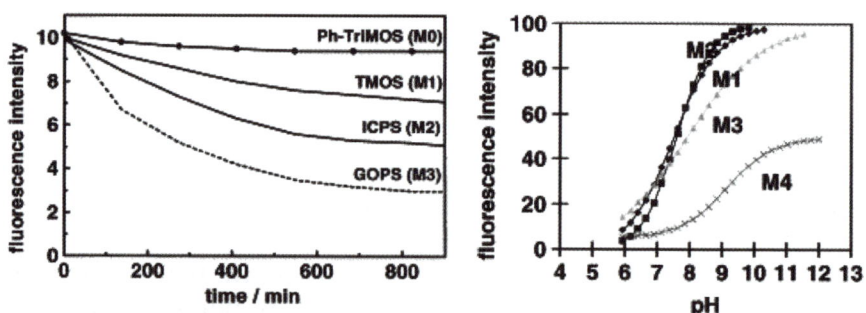

Figure 6.3 Leaching of sensor layers M4, M1, M2 and M3 (from top) on exposure to a flow of buffer solution (left) and titration plots of AF in poly-TMOS (M1), an organically modified silicate (M4), and covalently immobilized on ICPS (M2) and GOPS (M3) (right). (Reproduced from ref. 4, with permission.)

Deprotonation of AF in M4 is disfavoured because of the lipophilic microenvironment of the immobilized indicator, whereas the apparent pK_a of sol–gel immobilized dyes depends on the nature of the sol–gel matrix.

Once the leaching is prevented by suitable adjustment of the xerogel properties, the versatility of the sol–gel chemistry shows its full potential as shown, for example, in the development of a rapid (response rate of less than 5 s) and accurate (a pH precision of approximately 0.001 pH units) "off-the-shelf" detection fibre-optic pH sensor that can be applied to bioreactor monitoring, cell-culture monitoring, waste-water discharge monitoring, atmospheric/acid rain measurement and blood and body fluid measurements.[5] The sensor is based on measuring the ratio of absorbance of a pH indicator dye at *two* wavelengths. The ratiometric method is immune to changes in intensity and moderate changes in dye concentration due to leaching. Unlike other optical pH systems that measure fluorescence, this system is ratiometric and immune to drift. It offers very stable calibration (good for over a month), and can easily be incorporated into a variety of product packages, or applied to probes or as a coating to cuvettes, Petri dishes, flow cells or other media where the sample volume may be limited. Different pH indicators may be trapped in the sol–gel for use in different pH ranges. For example, bromocresol green (BCG) is responsive from pH = 5 to 9 (Figure 6.4), ideally suited for biological systems, environmental sampling and many industrial processes.

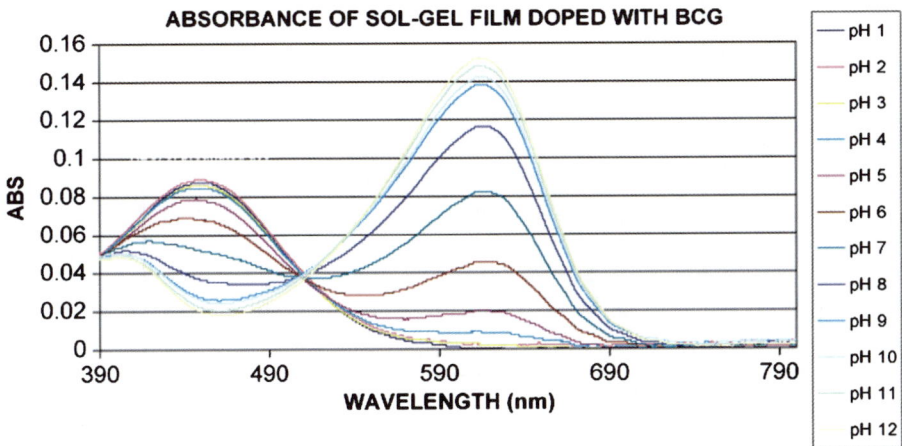

Figure 6.4 Plotting absorbance *versus* wavelength at several levels of pH for a BCG optical sensor shows that the alkaline form of the indicator absorbs at about 620 nm and the acidic form absorbs at about 450 nm. At about 550 nm both forms have the same molar absorptivity (isobestic wavelength). (Reproduced from LASERFOCUSWORLD.COM.)

6.2 Array Sensors

Xerogel sensors are extremely sensitive and offer unique advantages. Oxygen-sensing materials based on spin-coated *n*-propyltrimethoxysilane and 3,3,3-trifluoropropyltrimethoxysilane doped with the luminophore tris(4,7-diphenyl-1,10-phenathroline) ruthenium(II) ($[Ru(dpp)_3]^{2+}$) recently opened the way to a second generation of commercial oxygen optical sensors that are replacing the earliest electrochemical methods to detect this ubiquitous reactant (Figure 6.5).[6]

These sensors are based on the oxygen quenching of $[Ru(dpp)_3]^{2+}$ sequestered within the xerogels, and shows a stable response (constant to within 2%) over a six-month period (compared to the fivefold decrease in sensitivity in pure SiO_2-based xerogels). Moreover, the sensor exhibits a sensitivity significantly greater than observed with any previous $[Ru(dpp)_3]^{2+}$-based quenchometric sensor.

Scanning electron microscopy (SEM) used to investigate the structure of similar organically modified silicate (ORMOSIL) films shows that certain $[Ru(dpp)_3]^{2+}$-doped octyl-triethylorthosilicate (triEOS)–tetraethylorthosilicate (TEOS) composites form uniform, crack-free xerogel films (Figure 6.6) that can be used to construct high-sensitivity oxygen

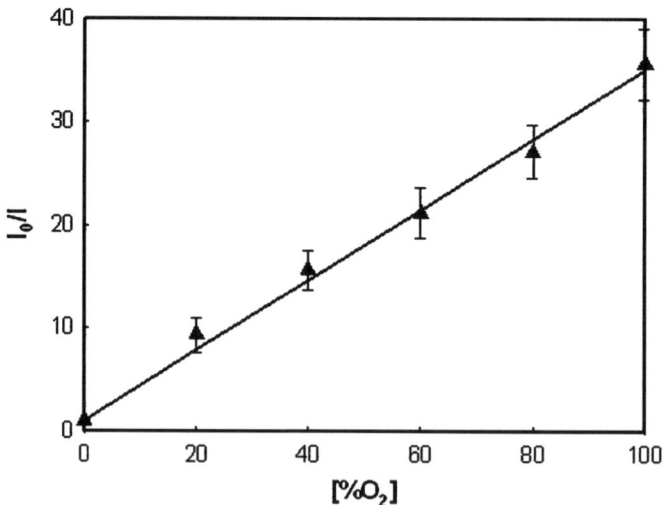

Figure 6.5 A fluorinated organically modified silicate doped with $[Ru(dpp)_3]^{2+}$ is a highly sensitive O_2 sensor. Fluorine here ensures unprecedented sensitivity and a remarkable stability (2% drift over 6 months). The material has been implemented in sol–gel handheld oxygen sensors that are already commercialized. (Reproduced from ref. 6, with permission.)

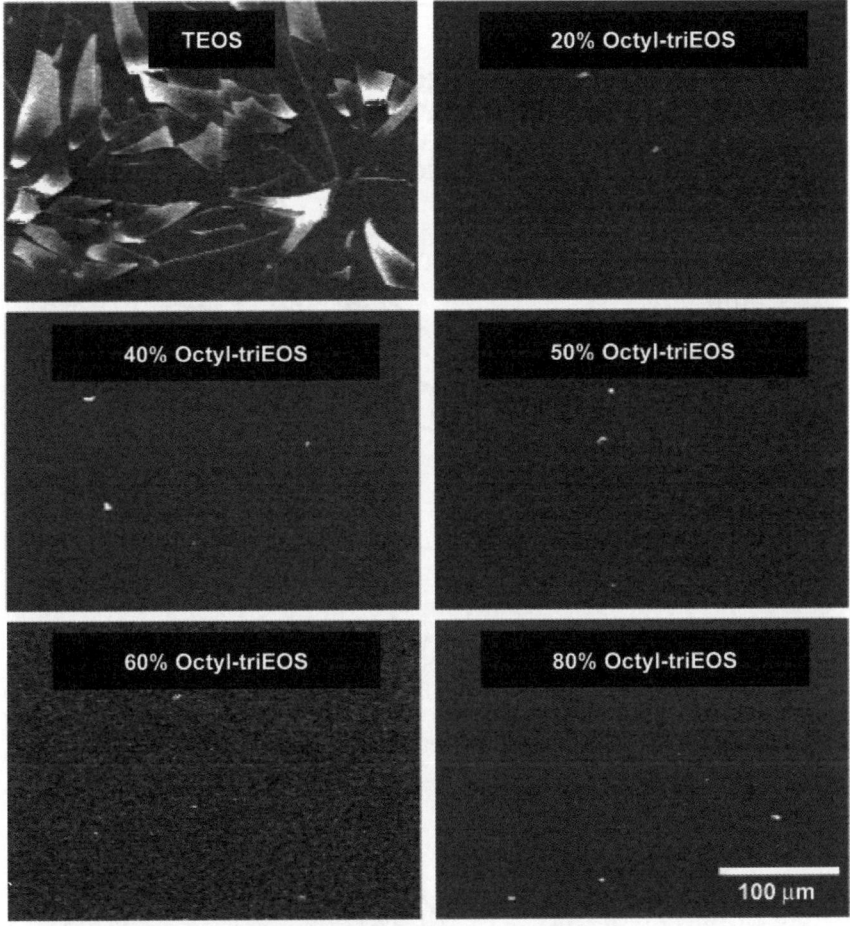

Figure 6.6 Low-resolution SEM images of xerogel films. The TEOS film is 2 months old; all other films are 3 months old. Film age does not affect the SEM images. (Reproduced from ref. 7, with permission.)

sensors that have linear calibration curves and excellent long-term stability.[7]

For example, an 11-month-old sensor based on 50 mol% octyl-triEOS exhibits more than fourfold greater sensitivity in comparison to an equivalent sensor based on pure TEOS (Figure 6.7). Over an 11-month time period, the sensitivity of a pure TEOS-based sensor drops by more than 400%, whereas a sensor based on 50 mol% octyl-triEOS remains stable (relative standard deviation *ca.* 4%). Integrated in fibre optics, optical sensors based on these materials are commercialized by the US

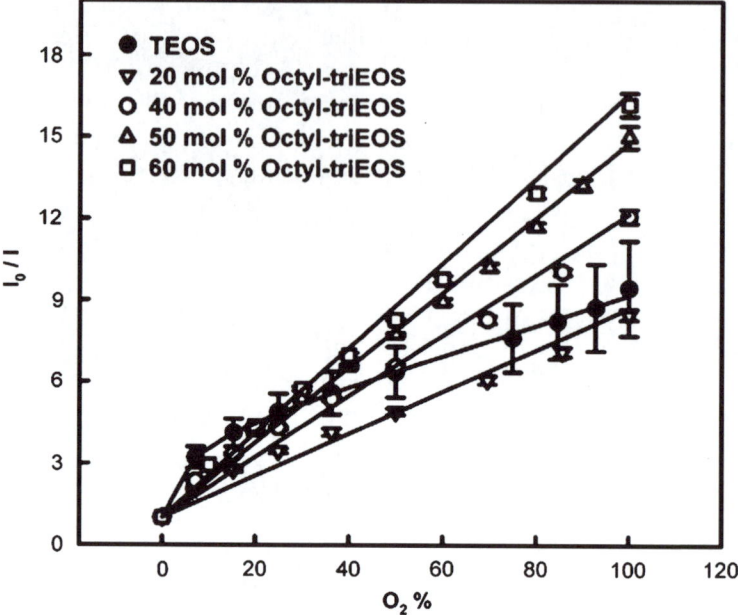

Figure 6.7 Typical intensity-based Stern–Volmer plots for 3-month-old $[Ru(dpp)3]^{2+}$-doped octyl-triEOS–TEOS composite xerogels. The solid lines represent the best fit to a Demas (TEOS) or Stern–Volmer model (all others). (Reproduced from ref. 7, with permission.)

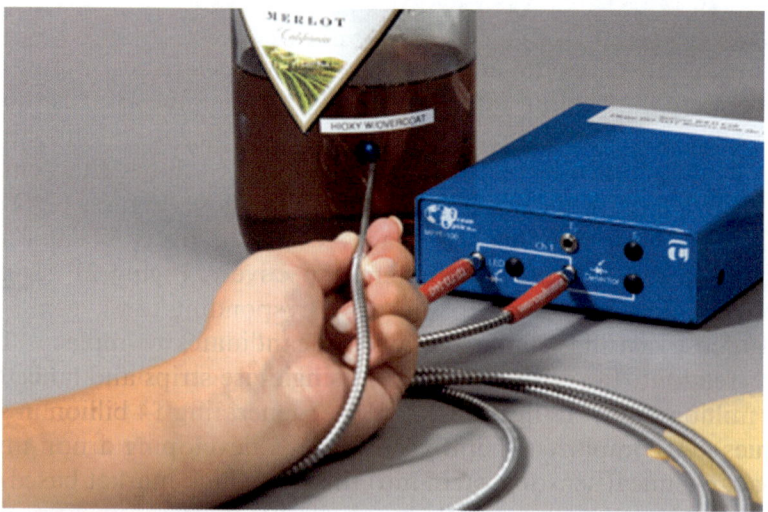

Figure 6.8 Labelled RedEye oxygen patch (Ocean Optics).

company Ocean Optics. One such sol–gel sensor is used to measure oxygen levels in packaging and other enclosed containers in medical, pharmaceutical, food and beverage applications. The patch (Figure 6.8) can be integrated into the surface of sample containers such as blood bags, pill blister packs or point-of-care analysis devices, to permit non-invasive, through-the-package oxygen concentration measurements.

Depending on the application, the simple presence of oxygen can be visually determined by colour change with a handheld light-emitting diode, or a fluorimeter can be used to directly measure the exact oxygen level. RedEye coatings are capable of monitoring low levels of oxygen in gas (to 0.005%) and dissolved oxygen in liquids (to 20 ppb), as well as the higher oxygen levels present in cell culture and respiratory monitoring.

Leach-proof sol–gel sensors can be arranged into solid-state, optical sensor arrays integrated with a light source (OSAILS) for the detection of multiple analytes,[8] particularly using pin-printed microarrays in which microspots are deposited via the sol–gel process. Since the early 1990s trapped enzymes have been used for making biosensors. These spots are three-dimensional allowing for higher loading than can be obtained from monolayers, and are ideally suited for protein-based biosensors. For example, an integrated light source/biosensor array is obtained by the combination of pin-printing and sol–gel processing techniques which provides a simple method to rapidly fabricate (<1s per sensor element) reusable sol–gel biosensor (SiO_2 doped with glucose oxidase) arrays that exhibit good analytical figures of merit.[9]

Figure 6.9 displays the array's response characteristics. Figures 6.9A and B present false colour fluorescence images when the array is subjected to air-saturated buffer that contains 10 mM glucose or air-saturated buffer without glucose, respectively. Figure 6.9C shows the composite calibration curve for the array responding to glucose. Figures 6.9D and E show false colour fluorescence images when the array is subjected to nitrogen- or oxygen-saturated buffer, respectively. Figure 2F shows the composite calibration curve for the array responding to oxygen.

In glucose sensing, similar arrays are candidates to replace "fingersticks" readouts (a reading device and single-use strips and lancets used by 15 million diabetics in the USA alone, generating $4 billion in sensor revenues). A company (TheraSyn) is indeed developing a non-invasive compact chemical sensor system into the fingerstick market based on an integrated photodetector coupled with a solid-state xerogel-based thin-film sensor (Figure 6.10).

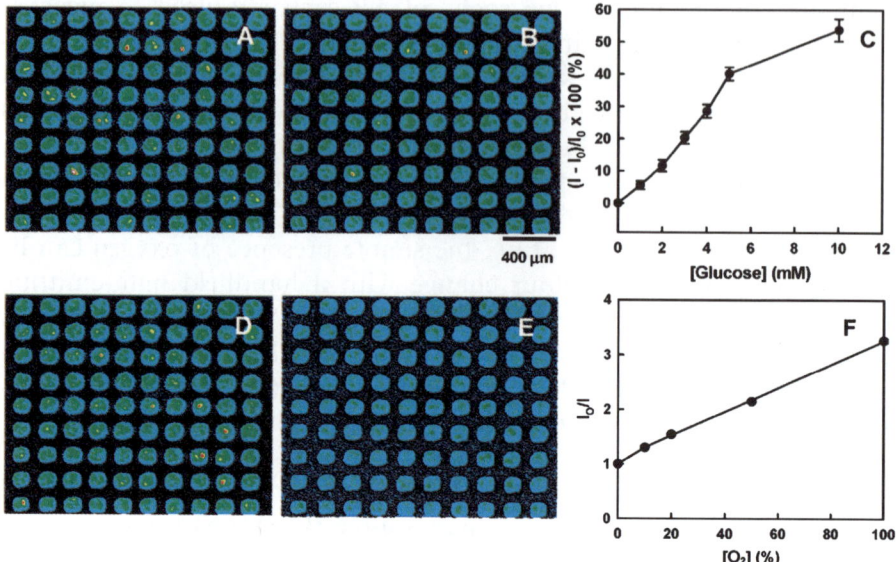

Figure 6.9   (A) Array challenged with air-saturated buffer that contains 10 mM glucose. (B) Array challenged with buffer alone, no glucose. (C) Typical composite glucose calibration curve for 100 elements. (D) Array challenged with nitrogen-saturated buffer. (E) Array challenged with oxygen-saturated buffer. (F) Typical composite oxygen calibration curve for 100 sensor elements. (Reproduced from ref. 9, with permission.)

In contrast to other tests that measure glucose, this method is truly non-invasive. The fluid for doing the test in fact is tears, as the glucose concentration in tears is known to correlate with the blood glucose concentration. Manufacturing is scheduled by 2010 with clinical trials currently being conducted and regulatory approval scheduled in 2009.

6.3   Templated Xerogels as Selective Chemical Sensors

Sol–gel sensors can be easily prepared to function as site selectively templated and tagged xerogels (SSTTXs). The resulting platform is completely self-contained, and it achieves analyte recognition without the use of biomolecules,[10] as shown for example by the selective detection and quantification of model compound 9-anthrol (Figure 6.11).

The results suggest SSTTXs as new sensor elements for glycoside, pharmaceutical, prostaglandin and steroid sensors. Indeed, a 2 μm thick film sensor (i) provides 0.3 μM detection limits for 9-anthrol; (ii) yields a 45 s response time; and (iii) is completely reversible (6% relative

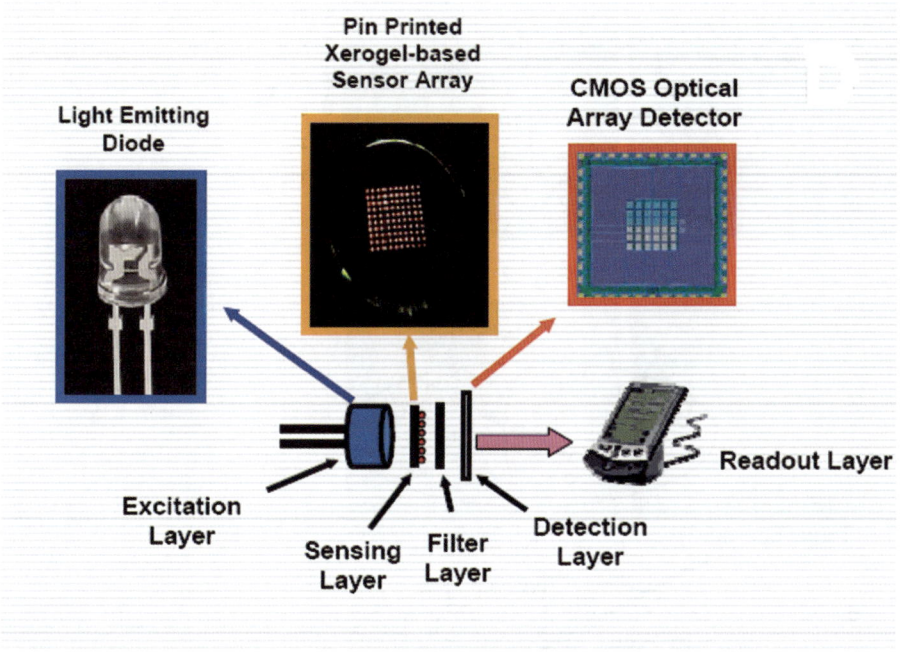

Figure 6.10 Non-invasive glucose sensor developed by TheraSyn Sensors Inc. will reach the market within the next two years. (Image courtesy of Prof. F. V. Bright.)

standard deviation after 25 cycles), yielding a selectivity factor for 9-anthrol of between 290 and 520 over several structurally similar analogues/interferences (*e.g.* anthracene, 9,10-anthracenediol, benzo-phenone, 2-naphthol, phenol and pyrene). The performance is stable (<2% drift in performance) for at least 10 months when the sensor is stored under ambient conditions in the dark.

The concept is extendable to templated sensors made of protein templated xerogels in which a luminescent reporter group is further added in close proximity to the template site so as to effectively transduce the protein–molecularly imprinted polymer binding event (Figure 6.12).[11]

The formulation of the ORMOSIL precursor is chosen in the light of screening experiments (Figure 6.13). For the recognition of ovalbumin, for example, the precursor is a 55 mol% TEOS, 2 mol% 3-aminopro-pyltriethoxysilane (APTES), 3 mol% *n*-octyltrimethoxysilane (C8-TMOS) and 40 mol% bis(2-hydroxyethyl)aminopropyltriethoxysilane (HAPTS) mixture. The resulting microarray sensor exhibits analytical

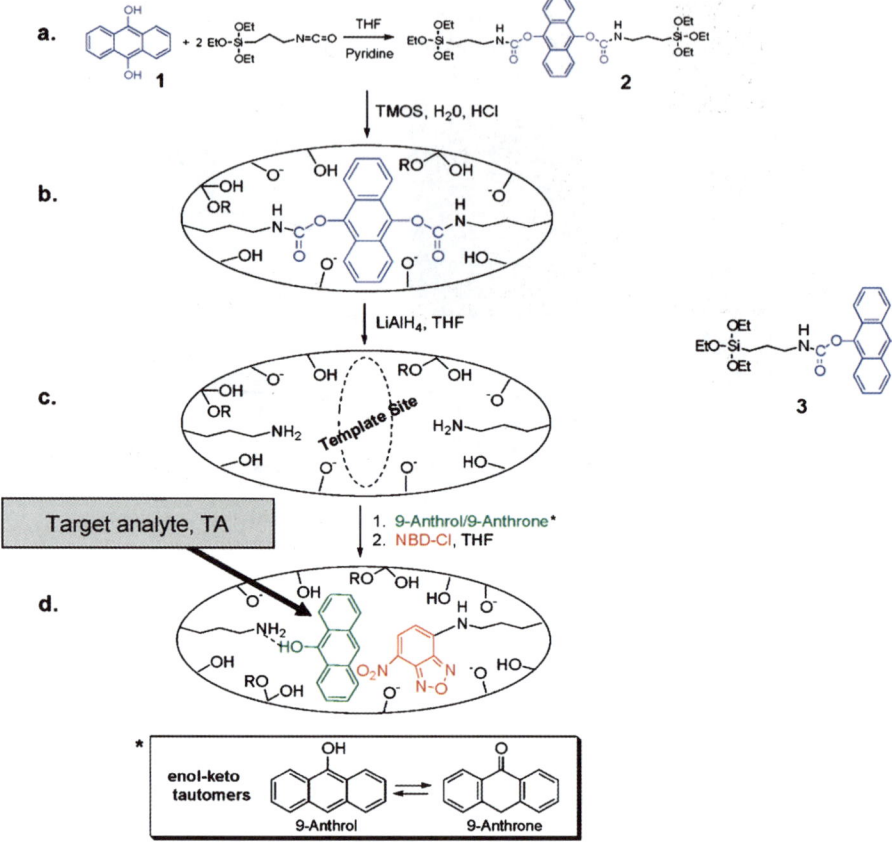

Figure 6.11 Reaction protocol for producing a SSTTX (the target analyte (TA) in this example is 9-anthrol). (Reproduced from ref. 10, with permission.)

figures of merit akin to an antibody-based assay, but the sensors are faster in comparison to an antibody-based assay, and the sensor elements are substantially more robust.

The concept of thin films of a molecularly imprinted sol–gel polymer with specific binding sites for a target analyte is general and can be applied also to electrochemical sensors. For example, a sensor to detect parathion in aqueous solutions is based on films cast on glass substrates and on glassy carbon electrodes (Figure 6.14).[12]

Again, the imprinted films show high selectivity towards parathion in comparison to similar organophosphates. The binding has been shown to be very sensitive to the type of functional monomer used for imprinting, and that rational design of the matrix components is an

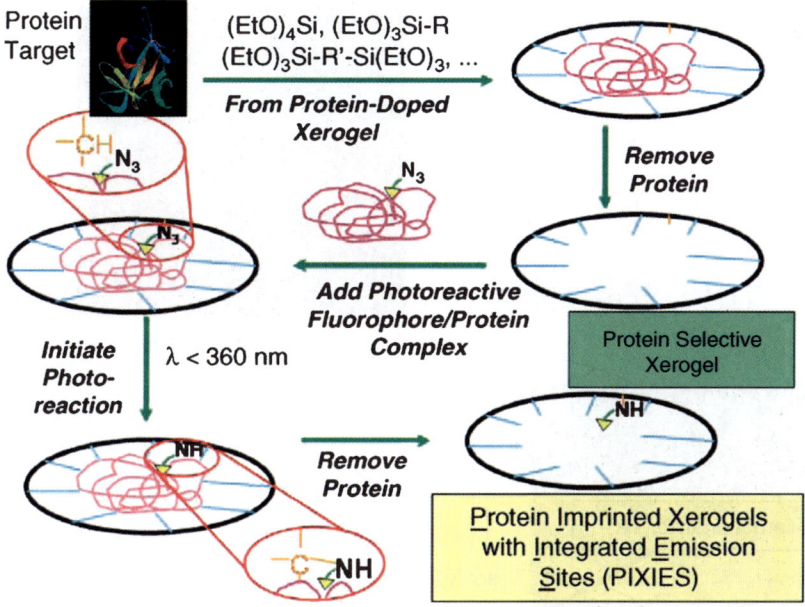

Figure 6.12 The PIXIES fabrication process. This is not an exhaustive listing of R and R' groups that one could use. Examples of alkoxides that have been tested are listed in the text. (Reproduced from ref. 11, with permission.)

essential step in molecular imprinting. The approach has been demonstrated also using dopamine (DA) as the template condensing phenyltrimethoxysilane (PTMOS), methyltrimethoxysilane (MTMOS) and TMOS in the presence of dopamine (Figure 6.14).[13] The electrode resulting from coating with such molecularly imprinted film is highly selective for DA.

6.4 Forthcoming Sensors

Whole cell–silicate sensors[14] will soon make use of the viability of cells and their ability to respond to external conditions, using genetically engineered cells tailored to emit measurable optical or electrical signals. The operation principle is shown in Figure 6.15.

Premkumar and co-workers demonstrated that sol–gel encapsulation meets all these requirements. In a series of articles they demonstrated successful encapsulation of recombinant *E. coli* which responded to different stress conditions by expression of fluorescent proteins (such as green and red fluorescent proteins) or bioluminescence in thin sol–gel

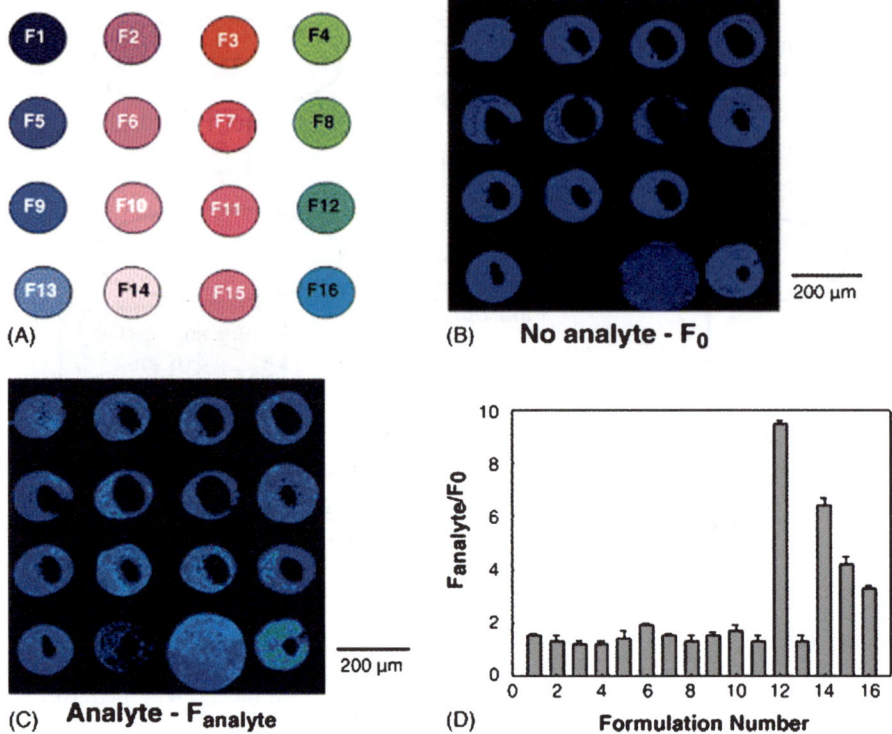

Figure 6.13 Screening results from 16 TEOS–APTES–C8-TMOS–HAPTS-based formulations F1–F16. (A) Formulation numbering scheme. (B) False colour CCD image from an array of PIXIES formulations in buffer ($\lambda_{ex} = 488$ nm, $\lambda_{em} > 500$ nm). (C) Same as (B) when challenged with 50 µM ovalbumin. (D) $F_{analyte}/F_0$ from (B) and (C). Formulation no. 12 appears to be the most analytically useful. (Reproduced from ref. 15, with permission.)

films[15] or glass beads[16] (Figure 6.15). The sensors maintained near-constant activity for several months, and no proliferation or bacterial die-off was observed during this period. The solid biohybrids allowed detection of general stress conditions, sensitive to a large range of chemicals which had similar physiological effects (*e.g.* genetic, oxidative or outer membrane damage) or to more specific inducers (hydrogen peroxide). Moreover, since the encapsulated bacteria did not proliferate, the researchers demonstrated multiple use of the same sensor. Based on the light response of the encapsulated bacteria after exposure to various luminescence inducers at different doses and the evolution of luminescence during the gelation process, it was concluded that the recombinant bacteria have a high biocompatibility with sol–gel derived silicate gels.

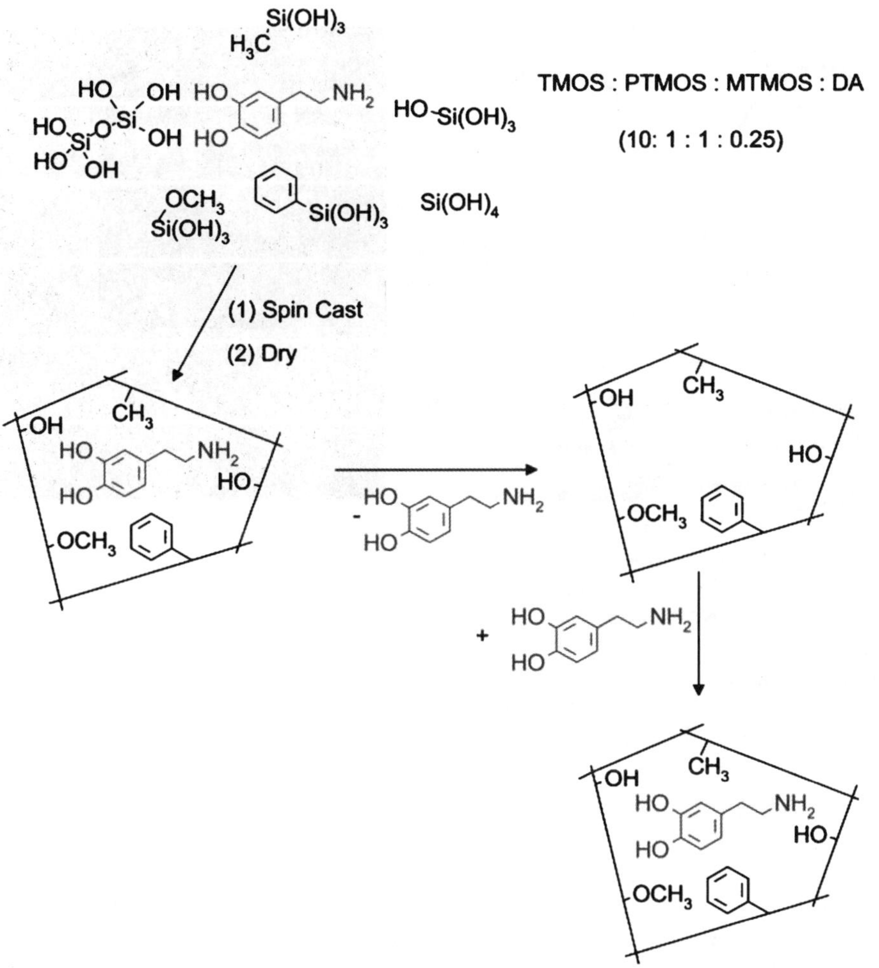

Figure 6.14 Simplified scheme for the sol–gel imprinting approach as it pertains to the templating of dopamine in functionalized silica. (Reproduced from ref. 14, with permission.)

Sensing of encapsulation-induced physiological stress will help in the devising of synthesis procedures that will minimize cell inactivation during immobilization.

Finally, we may expect to see doped silica sensors in which surfactants co-entrapped with the sensing molecule are used to tailor the properties of the dopant itself. This has been demonstrated by careful co-entrapment in a single sol–gel interphase of different surfactants along with a dopant dye (crystal violet; Chapter 1) as fluorescent pH sensor observing changes in the sensing properties of the resulting materials from large magnitudes to delicate fine-tuning.[17]

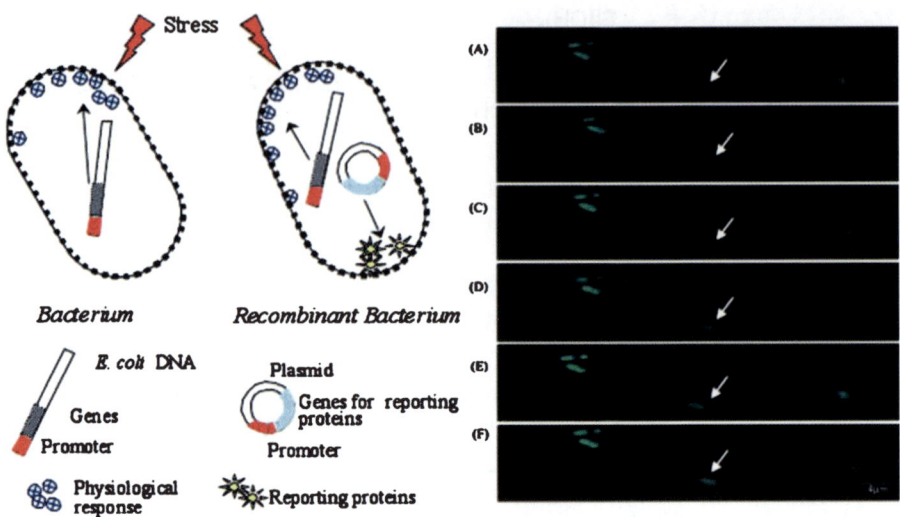

Figure 6.15   Left: stress-induced response of native and recombinant bacteria. Right: single *E. coli* cell fluorescent response after induction by mitomycin C. (Reproduced from ref. 16, with permission.)

References

1. D. Avnir, L. C. Klein, D. Levy, U. Schubert and A. B. Wojcik, Organo-silica sol-gel materials, in *Organic Silicon Compounds*, ed. Z. Rappoport and Y. Apeloig, John Wiley & Sons, 1998, vol. 2.
2. M. R. Shahriari, Sol-gel fiber optic chemical sensors, in *Optical Fiber Sensor Technology*, ed. K. T. V. Grattan and B. T. Meggitt, Kluwer Academic, Dordrecht, 1999, vol. **4**, ch. 3.
3. R. Zusman, C. Rottman, M. Ottolenghi and D. Avnir, Doped sol-gel glasses as chemical sensors, *J. Non-Cryst. Solids*, 1990, **122**, 107.
4. A. Lobnik, I. Oehme, I. Murkovic and O. S. Wolfbeis, pH optical sensors based on sol–gels: chemical doping versus covalent immobilization, *Anal. Chim. Acta*, 1998, **367**, 159.
5. U. H. Manyam, Application of dye-doped sol gel for pH sensing, PhD thesis, Rutgers University, 2006.
6. R. M. Bukowski, R. Ciriminna, M. Pagliaro and F. V. Bright, *Anal. Chem.*, 2005, 77, 2670. For instance, a new easy-to-use handheld optical sol–gel oxygen sensor (named FOXY-LITE, see also the website: http://www.oceanoptics.com) with a response time of less than one second is commercialized at $1999 since March 2004 by the US company Ocean Optics Ltd.

7. Y. Tang, E. C. Tehan, Z. Tao and F. V. Bright, Sol-gel-derived sensor materials that yield linear calibration plots high sensitivity, and long-term stability, *Anal. Chem.*, 2003, **75**, 2407.
8. E. J. Cho and F. V. Bright, Optical sensor array and integrated light source, *Anal. Chem.*, 2001, **73**, 3289.
9. E. J. Cho, Z. Tao, E. C. Tehan and F. V. Bright, Multianalyte pin-printed biosensor arrays based on protein-doped xerogels, *Anal. Chem.*, 2002, **74**, 6177.
10. E. L. Shughart, K. Ahsan, M. R. Detty and F. V. Bright, Site selectively templated and tagged xerogels for chemical sensors, *Anal. Chem.*, 2006, **78**, 3165.
11. Z. Tao, E. C. Tehan, R. M. Bukowski, Y. Tang, E. L. Shughart, W. G. Holthoff, A. N. Cartwright, A. H. Titus and F. V. Bright, Templated xerogels as platforms for biomolecule-less biomolecule sensors, *Anal. Chim. Acta*, 2006, **564**, 59.
12. S. Marx, A. Zaltsman, I. Turyan and D. Mandler, Parathion sensor based on molecularly imprinted sol-gel films, *Anal. Chem.*, 2004, **76**, 120.
13. R. Makote and M. M. Collinson, Template recognition in inor-ganic-organic hybrid films prepared by the sol-gel process, *Chem. Mater.*, 1998, **10**, 2440.
14. D. Avnir, T. Coradin, O. Lev and J. Livage, Recent bio-applications of sol–gel materials, *J. Mater. Chem.*, 2006, **16**, 1013.
15. J. R. Premkumar, O. Lev, R. Rosen and S. Belkin, Encapsulation of luminous recombinant *E. coli* in sol-gel silicate films, *Adv. Mater.*, 2001, **13**, 1773.
16. M. L. Ferrer, L. Yuste, F. Rojo and F. del Monte, A biocompatible sol-gel route for encapsulation of living bacteria in organically modified silica matrixes, *Chem. Mater.*, 2003, **15**, 3614.
17. C. Rottmann and D. Avnir, Getting a library of activities from a single compound: tunability and very large shifts induced by sol-gel entrapped micelles, *J. Am. Chem. Soc.*, 2001, **123**, 5730.

Hybrid Silica-Based Composites

7.1 Hybrid Polymer–Silica Composites

Hybrid polymer–silica nanocomposites formed from various combinations of silicon alkoxides and polymers to create a nanoscale admixture of silica and organic polymers constitute a class of composite materials with combined properties of polymers and ceramics. They are finding increasing applications in protective coatings (Figure 7.1), optical devices, photonics, sensors and catalysis.[1]

Typically, organic groups chemically linked to the alkoxides are homogeneously incorporated using alkoxides of general formula $R'_n Si(OR)_{4-n}$, where the organofunctional group R' can react with itself or additional components (R' contains vinyl, methacryl or epoxy groups, for example). The organic groups bring about a flexible network with increased relaxation properties, and the densification temperature can be drastically reduced in comparison to pure inorganic sol–gel materials. Hybrids composed of inorganic oxides covalently bonded to organic polymers are of particular interest, since they benefit from the lack of interface imperfections and because the properties of the resulting materials are intermediate between those of polymers and glasses, material properties that are simply not afforded by organic polymers and glasses (Figure 7.2).[2]

The hybrids generated by the sol–gel process combine the flexibility and mechanical strength of the organic constituent with the hardness, stiffness and transparency of the inorganic silica network. Hardness is often further conferred by the employment of other inorganic oxide particles (such as alumina or titania). This first started in the 1980s when

Silica-Based Materials for Advanced Chemical Applications
By Mario Pagliaro
Published by the Royal Society of Chemistry, www.rsc.org

Figure 7.1 HSG has pioneered the commercialization of ultraviolet-curable hybrid glass protective coatings.

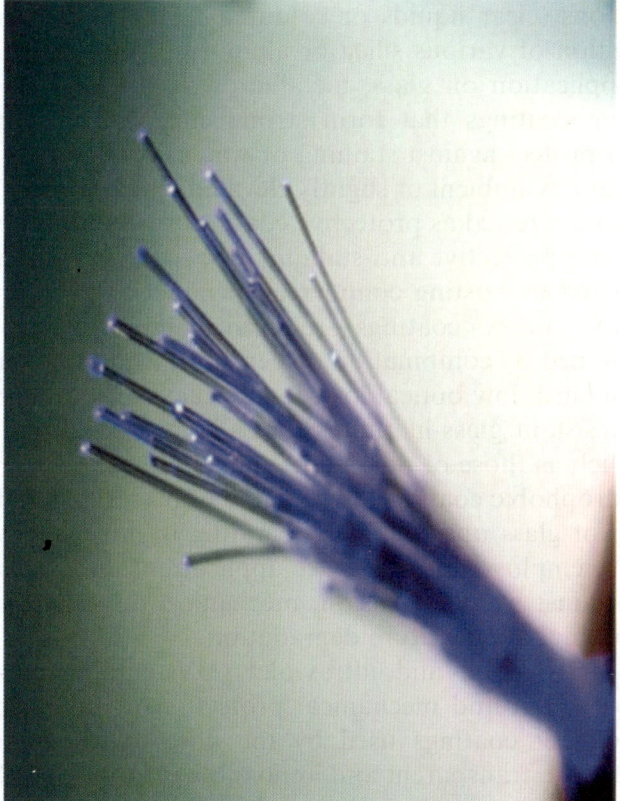

Figure 7.2 Hybrid sol–gel coatings are ideally suited to protect optical fibres.

researchers at the Institut fur Silicatforschung in Germany developed the "chemically controlled condensation" method, namely hydrolytic polycondensation using water generated *in situ*, which is the basis for making ORMOCER materials commercialized as protective hard coatings for transparent plastics.

Further examples soon followed such as in photochromic glasses, contact lenses, solid-state lasers and waveguides and as replacements for silicon dioxide as insulating materials in the microelectronics industry. Hybrid polymers, for example, are used to manufacture contact lenses because of the versatile tailoring permitted by hybrid materials chemistry. A silicon-like unit such as $-Si(CH_2)_n-O-CO-CH(CH_3)=CH_2$ chosen to favour oxygen diffusion is copolymerized with hydroxyethyl methacrylate (HEMA) in the presence of $Ti(OR)_4$ as condensation catalyst. The resulting hybrid material exhibits good oxygen permeation (to maintain the oxygen supply to the cornea), good mechanical properties and wettability along with enhanced scratch resistance.[3]

Formulations (clear liquids or coloured formulations) based on an ethanol solution of various silica or aluminosilica precursors are marketed for application on glass, metallic or plastic surfaces to achieve hard 1–10 μm coatings that form strong chemical bonding with the surfaces and protect against staining or water corrosion.

Curing requires ambient or slightly elevated temperatures. The formulations have been tested as protective coatings for sandblasted glass and exhibit superior protective anti-staining properties and water repellency when compared to existing commercial products. Others are excellent low refractive index coatings—refractive indices from 1.3800 to 1.4500—obtained by combination of perfluorinated polymer and silica, as they form hard, low optical loss, low refractive index, thermally and chemically resistant glass-like films that adhere well to glass or plastic substrates such as those of polymeric optical fibres.

Other hydrophobic coatings are capable of preserving and enhancing the strength of glass optical fibres operating in adverse environments, such as those employed in aerospace applications, as the coatings protect fibres against water corrosion, mechanical and chemical damage, abrasion and high-temperature degradation.

Such sol–gel derived and ultraviolet (UV)-curable hybrids afford enhanced corrosion and mechanical protection of fibres compared to standard polymer coatings used by the fibre optics industry. These coatings are hard, transparent and impossible to strip since they become intimately bonded to the fibre surface during curing.

These hybrid coatings are therefore more resistant to abrasion than polymer coatings. A short UV exposure cures not only the polymeric component but also enhances condensation of the inorganic component as well, because of exposure to the heat of the UV chamber rather than the UV radiation itself. Curing, in turn, leads to permanent bonding of the coating material to the fibre surface. Thinner coatings tend to exhibit higher strength and stress corrosion parameter values. If the content of polymerizable acrylate groups in a coating is reduced below a certain threshold (20% lower, for instance, than in optimal coatings), this

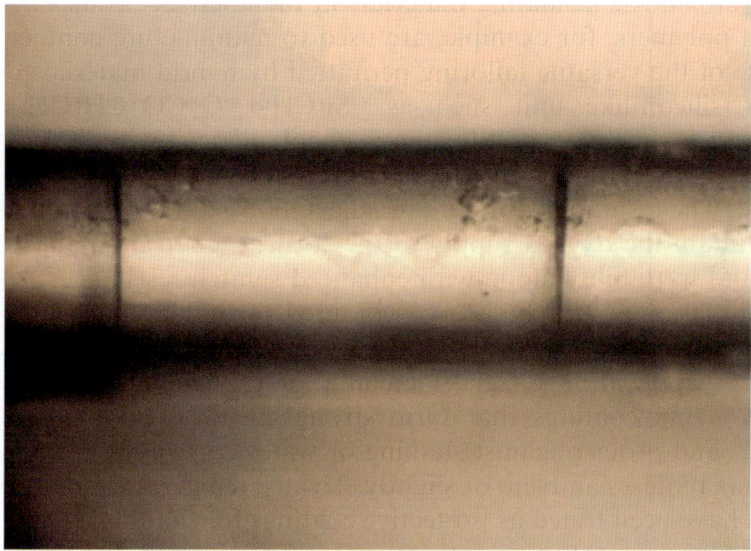

Figure 7.3 Cracks appear in the coating of optical fibres if the organic polymer content is reduced below a certain threshold. (Reproduced from ref. 4, with permission.)

induces stress in the solidifying network that causes its rupture when the whole inorganic–organic backbone goes through a relaxation process after curing, due to different ratios of cross-linking of both the inorganic and organic coating components (Figure 7.3).[4]

7.2 New Aqueous Route to Hybrid Silicas

A new aqueous sol–gel process for the production of hybrids has been recently developed that is capable of producing transparent organic–inorganic hybrid materials with desirable mechanical properties for several applications.[5] Contrary to the conventional wisdom that water should be strictly controlled in similar systems, here a large amount of water is used as solvent added to a mixture of an epoxy organofunctional silane with an aluminium alkoxide. Typically, 3-glycidyloxypropyltrimethoxysilane (GPTMS) and aluminium *sec*-butoxide (Al(OBu)$_3$; 80 wt% in isopropanol) are mixed at room temperature in Si:Al molar amounts ranging from 0.5 to 3 in an open container. Water (molar ratio H$_2$O:Al(OBu)$_3$ = 30:1) is then added into the mixture, with a precipitate forming immediately. The mixture is then heated to and maintained at 60 °C under vigorous stirring which results in precipitate peptization into a transparent solution.

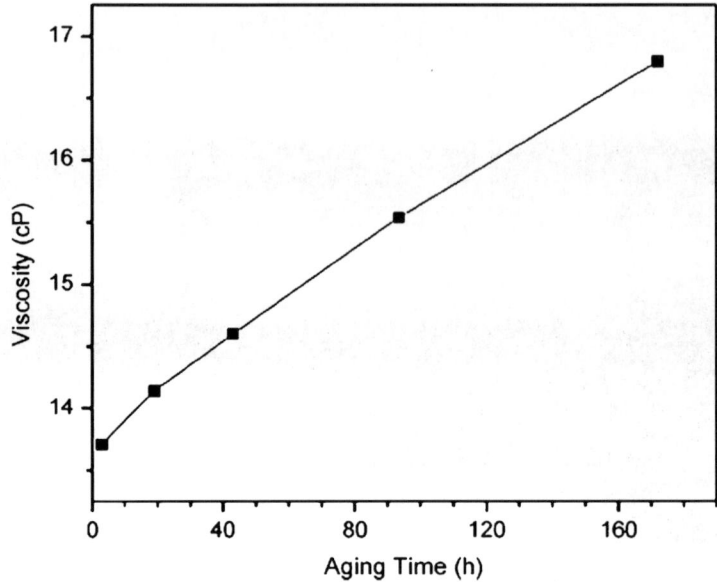

Figure 7.4 Viscosity as a function of ageing time of a GPTMS–Al(OBu)₃ hybrid solution at room temperature. GPTMS:Al(OBu)₃ = 3:2 (molar ratio). (Reproduced from ref. 5, with permission.)

The GPTMS facilitates the dissolution of the aluminium hydrates, as GPTMS is hydrolysed and forms condensates with the aluminium species in the solution. The mechanisms for the dissolution of the aluminium hydrates in the presence of GPTMS are not clear, but the peptization time increases with the GPTMS:Al(OBu)₃ molar ratio, with a critical amount of GPTMS (GPTMS:Al(OBu)₃ > 0.5) that is needed to peptize the system. The viscosity of the precursor solution increases by 20% over a period of a week (Figure 7.4) pointing to gel structures being developed in the precursor solution.

Comparison between the new homogeneous hybrid system and a nanoparticulate GPTMS–boehmite sol (with Si:Al molar ratio = 2:1 for both solutions) shows (Figure 7.5) that, despite a similar elemental constitution of the two systems, the nanoparticulate GPTMS–boehmite sol is translucent, whereas the GPTMS–Al(OBu)₃ hybrid system is highly transparent with no light scattering. This suggests that the size of the nanoparticles in the GPTMS–Al(OBu)₃ hybrid solution is indeed much smaller than that in the GPTMS–boehmite sol.

Accordingly, the Young's moduli for gel samples derived from the latter hybrid solution by our new sol–gel route were significantly higher

Figure 7.5 Image of two precursor solutions prepared using different methods: (a) simple mixing of GPTMS and a Yoldas sol; (b) mixing of GPTMS and Al(OBu)$_3$ followed by peptization in water. (Reproduced from ref. 5, with permission.)

than those for GPTMS–boehmite nanocomposite gels (Figure 7.6), showing evidence of a much denser structure than the nanoparticulate GPTMS–boehmite sol (the modulus increased significantly after being aged at room temperature for both samples due to condensation reactions, syneresis and progressive strengthening and stiffening of the gel matrix). The 3.5 GPa modulus of the GPTMS–Al(OBu)$_3$ hybrid aged gel is greater than those of most polymers and, most importantly, it can be tuned by changing the GPTMS:Al(OBu)$_3$ molar ratio in the precursor solution.

Easily applied onto various substrates by standard coating techniques, these hybrid coatings can be used as abrasion and scratch resistant coatings for plastic substrates. The results of Taber abrasion tests for various hybrid coatings on poly(methyl methacrylate) (PMMA) substrates in comparison to uncoated PMMA (Figure 7.7) show that after 500 cycles of the Taber abrasion test, the GPTMS–Al(OBu)$_3$ hybrid coatings produced by the new sol–gel route show haze values of 6–10%, depending on the GPTMS:Al(OBu)$_3$ molar ratio, whereas the uncoated PMMA shows a haze value of about 30%. This is a distinct improvement of abrasion resistance. All the coatings were cured at 80 °C, a low curing temperature, and the abrasion resistance of the hybrid coatings is comparable to that of ORMOCER-type coatings (the ORMOCER type and commercial siloxane hard coatings show a haze of about 6–15% after 500 cycles of the Taber abrasion test and are typically cured at 90–130 °C).

The aluminium component (aluminium *sec*-butoxide) in the hybrid gel catalyses the ring-opening reaction of the epoxy group in the GPTMS,

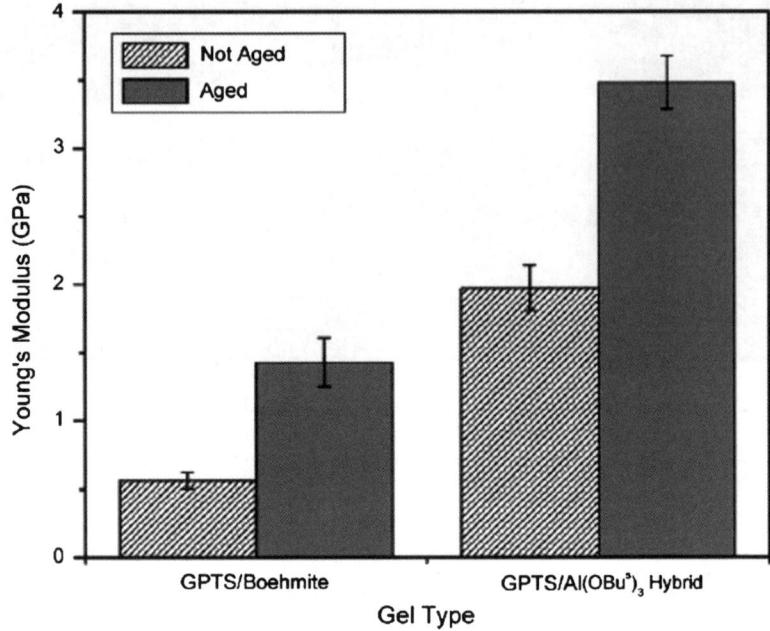

Figure 7.6 Young's moduli of the GPTMS–boehmite nanocomposite and the hybrid
gel derived from GPTMS–Al(OBu)$_3$ aqueous solution. Samples without
ageing were dried at 50 °C for 16 h and then at 110 °C for 16 h. The aged
samples were dried at room temperature for 9 days before being dried at
elevated temperatures (50 °C for 16 h and then 110 °C for 16 h). (Reproduced from ref. 5, with permission.)

possibly leading to a high degree of cross-linking of the gel sample. Indeed,
aluminium butoxyethoxide can catalyse epoxide ring opening as well as
inorganic condensation reactions (Scheme 7.1). The Fourier transform
infrared (FTIR) spectrum of the GPTMS–Al(OBu)$_3$ hybrid gel clearly
shows (Figure 7.8) two characteristic bands of the epoxides (at 950–810
and 1240–1260 cm^{-1}), with the former band shrinking significantly, and
the latter band disappearing, with the addition of Al(OBu)$_3$.

The aircraft manufacturer Boeing sponsored the work described here,
and the company is developing the technology further for aircraft
window coatings.

7.3   Hybrid Silica–Polymer Aerogels

Hybrid polymer–silica aerogels have exceptional insulating and mechanical properties and will soon be commercialized as high-performance,
low-cost insulators.[6] Inorganic–organic silica areogels are non-flammable

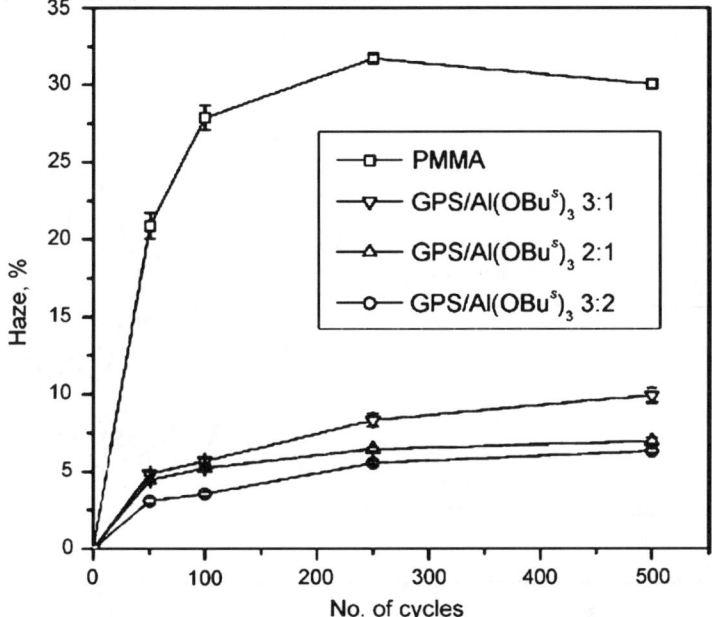

Figure 7.7 Abrasion resistance (haze) dependence on GPS:Al(OBu)$_3$ molar ratio for hybrid-coated PMMA surfaces compared to uncoated PMMA. PMMA substrate is corona treated followed by a 2.5% 3-aminopropyltriethoxysilane treatment prior to the application of GPS–Al(OBu)$_3$ hybrid solutions. Coating thickness: 4 mm; coatings were cured at 80 °C for 4 h. (Reproduced from ref. 5, with permission.)

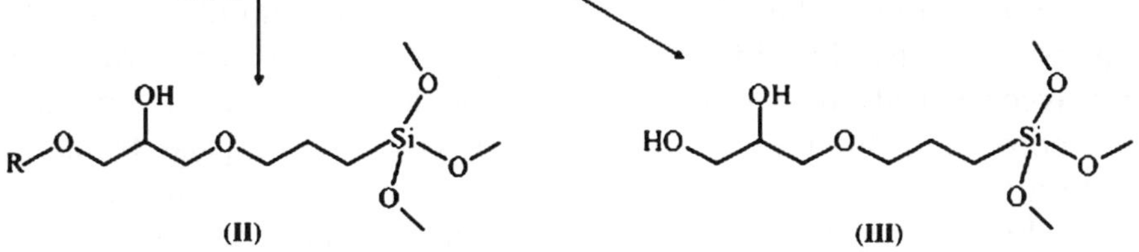

Scheme 7.1 Reaction pathways of GPTMS. The opened epoxide ring is converted to poly(ethylene oxide) (I), alkyl ether (II) and diol (III). (Reproduced from ref. 5, with permission.)

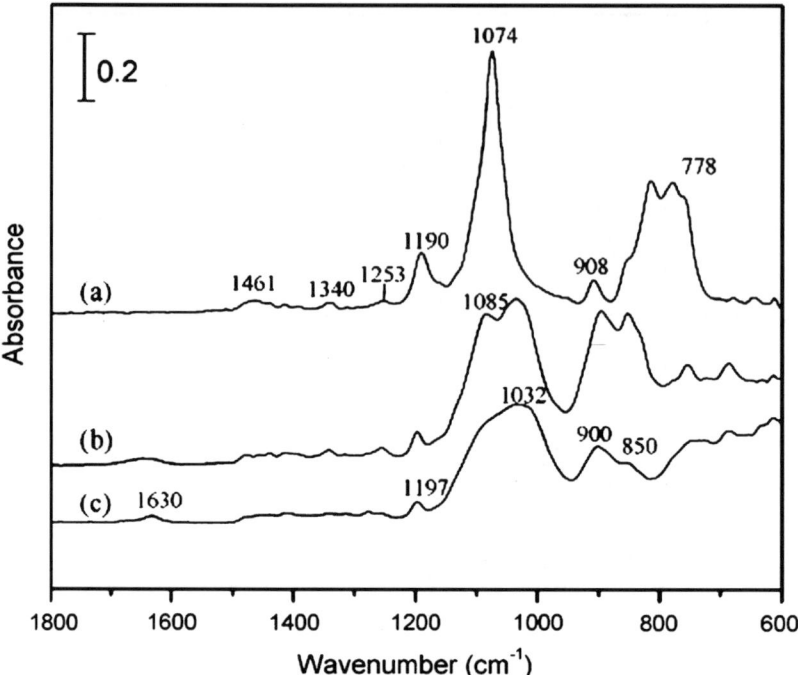

Figure 7.8   FTIR spectra of (a) GPTMS liquid control, (b) GPTMS gel and (c) GPTMS–Al(OBu)$_3$ (2:1) hybrid gel. The gel and hybrid gel samples were obtained by drying the precursor solutions at room temperature for 6 days. (Reproduced from ref. 5, with permission.)

and extremely light (densities in the range 3–500 kg m^{-3}) and are excellent thermal and acoustic insulators (thermal conductivity in the range 0.01–0.02 W m^{-1} K^{-1} and acoustic impedance between 103 and 106 kg m^{-2} s^{-1}). Traditional drawbacks such as high production costs, brittleness and instability towards atmospheric moisture are solved at the same time by a new sol route which will make aerogels the insulators of the future (Scheme 7.2).[7]

To overcome the fragility of the gels and the possibility of them undergoing fatigue when submitted to cycles of mechanical efforts, the synthesis of the hybrid wet gels follows a two-step hydrolysis/ co-polycondensation of tetraethylorthosilicate (TEOS) with functionalized core–shell polymer nanoparticles (PNPs) of low glass transition temperature (T_g), with excess water, in 2-propanol. The resulting nanohybrid aerogels are opaque monoliths that are stable under atmospheric conditions.

It is evident that the pore network is modified only by the presence of PNPs in the structure, since the light dispersion becomes so

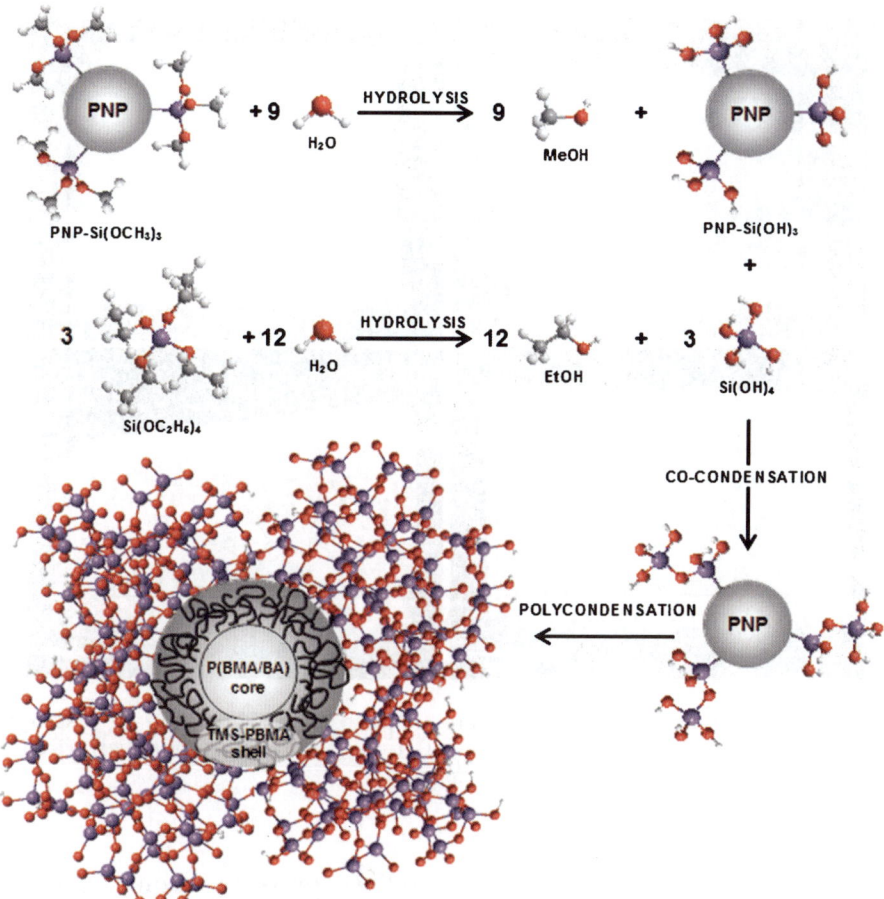

Scheme 7.2 Reaction scheme to obtain SiO₂–polymer nanoparticle (PNP) nanohybrids. (Reproduced from ref. 6, with permission.)

altered: whereas the dried inorganic sample is translucent, the hybrid samples are opaque even for very low polymer contents (Figure 7.9).

The PNPs contain a low-T_g trimethoxysilyl-modified poly(butyl methacrylate) (TMS-PBMA) shell and an even lower T_g poly[(butyl methacrylate)-*co*-(butyl acrylate)] (P(BMA–BA)) core. Their cross-linking stabilizes the spherical shape of the PNPs, allowing them to be dispersed in the sol–gel organic medium. The hydrolysed TMS groups at the shell surface anchor the nanoparticles to the silica network, and the core–shell low T_g values contribute to improve the mechanical properties of the hybrid material with respect to the inorganic SiO₂ aerogel. Chemical binding of PNPs to the growing silica network, in its turn,

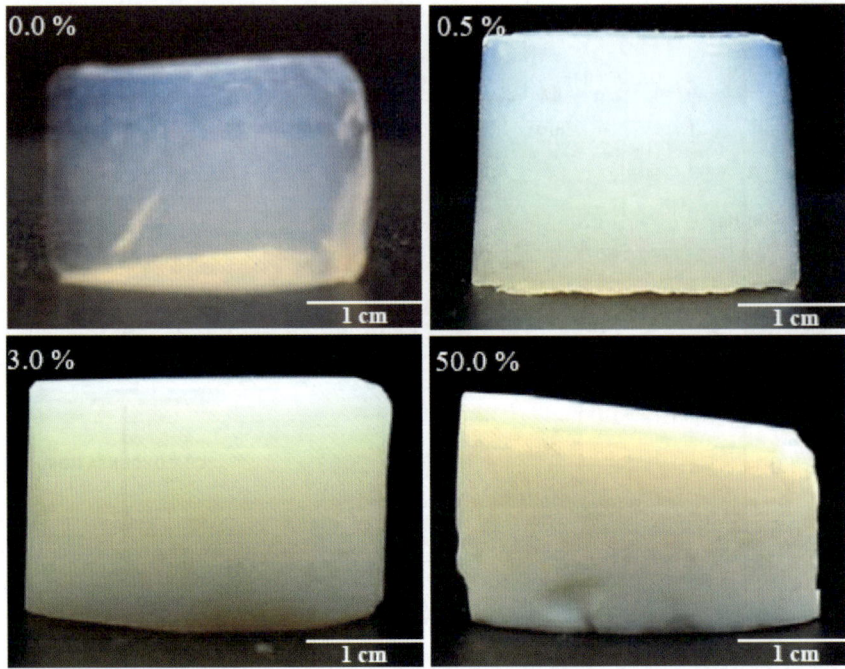

Figure 7.9 Photographs of slices of pure SiO$_2$ and three hybrid SiO$_2$–PNP aerogels with different polymer contents (weight percent as indicated). (Reproduced from ref. 6, with permission.)

allows the material to withstand capillary pressure upon subcritical drying.

The nanohybrid aerogels show improved mechanical properties (Figure 7.10) with respect to the corresponding inorganic aerogel (much higher energy is stored until fracture), without loss of structure or porosity. For example, the hybrid aerogel with 3 wt% of polymer has a density of 357 kg m^{-3}, total porosity of 83%, average mesopore diameter of 11.5 nm, mechanical resistance to compression of 4.24 MPa and maximum deformation of 14.4%. Such an amazing combination of properties makes these hybrid materials strong candidates for high-performance insulation applications.

The brittle behaviour of the silica aerogel remains upon the introduction of TMS-modified PNPs. However, even a low polymer content promotes a clear improvement of the mechanical properties of the material: Young's modulus increases and the maximum compression strength and the corresponding strain become three to five times higher. Furthermore, the improvement in mechanical behaviour is noticeable

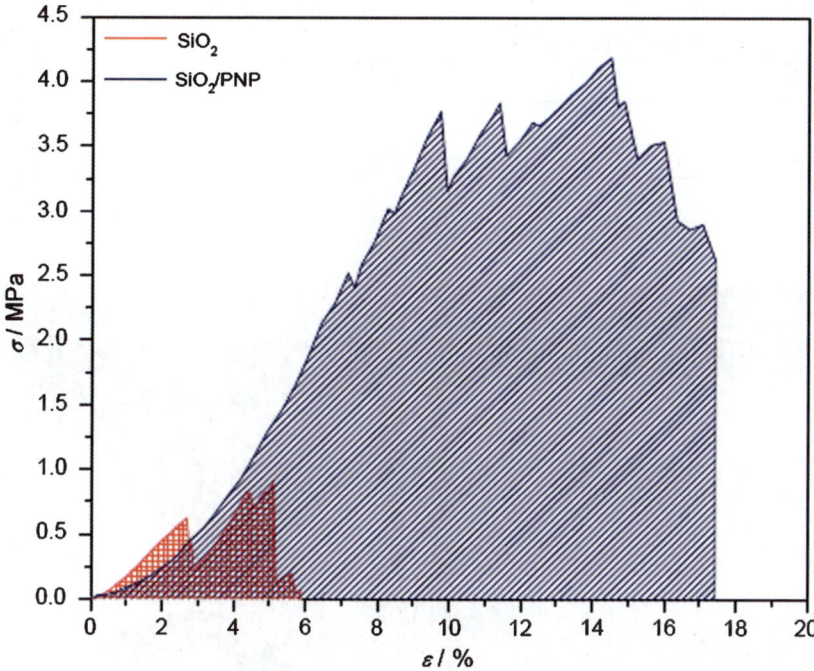

Figure 7.10 Stress–strain (σ–ε) curves obtained from unidirectional compression tests of pure silica and a hybrid aerogel containing 3 wt% PNPs. (Reproduced from ref. 6, with permission.)

from the much higher energy that the hybrid aerogel is able to absorb up to the maximum compression strength (roughly the area under the σ–ε curve).

The plots in Figure 7.10 show the typical features of brittle, cellular materials. The low-strain region corresponds to the reversible bending of cell walls, and the region with average lower slope relates to cell wall buckling. The serrations along the compression curve result from successive fracture of the cell walls. The region corresponding to general buckling and densification, where a re-increase in slope would be observed, is absent. The values of Young's modulus (E), the maximum compression strength (σ_{max}) and the corresponding strain ($\varepsilon_{\text{"max"}}$) obtained from these curves are summarized in Table 7.1.

7.4 Polyethylene@Silica

The first physically interpenetrating nanocomposite between polyethylene (PE) and silica (the simplest and most common organic and

Table 7.1 Compression properties of pure silica and hybrid aerogels containing 3 wt% PNPs.

Sample	E (MPa)	σ_{max} (MPa)	ε_{max} (%)
SiO_2	28	0.92	5.1
SiO_2–PNP (3 wt%)	44	4.24	14.4

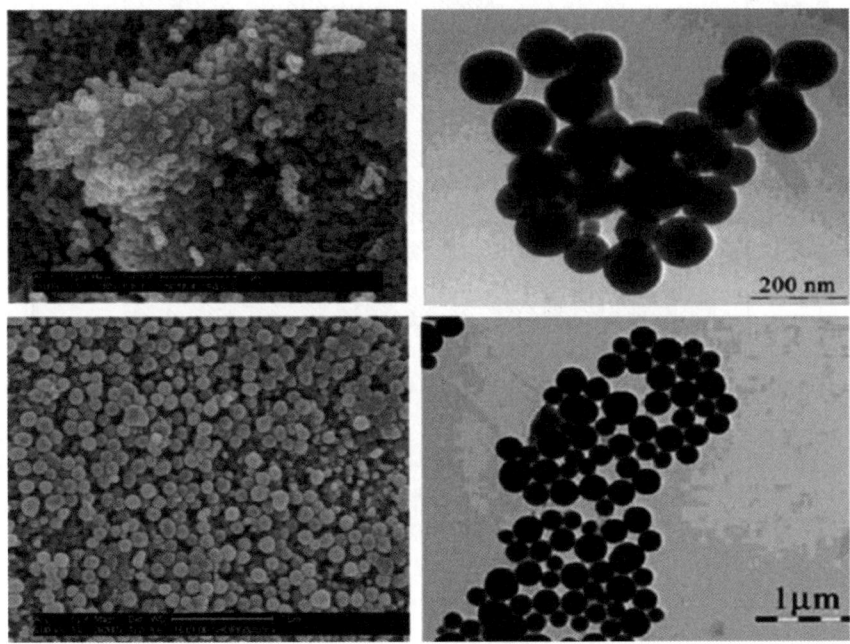

Figure 7.11 Scanning electron microscopy (left) and transmission electron microscopy (right) images of low-density PE@silica (top) and high-density PE@silica (bottom). (Reproduced from ref. 8, with permission.)

inorganic polymers) has been recently reported (Figure 7.11).[8] Preliminary studies point to an important application of the PE@silica particles. This is linked to their ability to disperse homogeneously in polyolefins due to the PE chains on the surface of the particles (based on "like dissolves like"), as well as successful compounding of mixtures of PE@silica and polyolefins.

Instead of blending them either by dispersing silica powders as fillers within PE, which is a well-developed technology with many applications, or by polymerizing ethylene within the pores of modified[9] or unmodified silica,[10] now the nanometric physical blending of PE is realized by the entrapment of dissolved PE in a polymerizing TEOS.

The key to the successful preparation of this new composite is the identification of a surfactant, PE-b-PEG, that is capable of stabilizing the emulsion and promoting the dissolution of the PE. Then submicrometre particles of low-density PE@silica and high-density PE@silica are synthesized by carrying out a silica sol–gel polycondensation process within emulsion droplets of TEOS-dissolved PE, at elevated temperatures (78 and 130 °C for low- and high-density PE, respectively).

The mechanism suggested for the formation of the particles as well as their inner structure involves three stages (Scheme 7.3). The first stage is the drop formation step when the surfactant facilitates the dissolution process of the polymer and stabilizes the forming emulsion (the "like dissolves like" principle operates here with the PE portion of the PE-b-PEG). Stabilization of the oily droplet is due to the two portions of the surfactant, each of which is very compatible with one of the two phases: PEG with the water–ethanol phase, and PE with the TEOS–xylene–PE phase in which the PE chains are fairly stretched.

The second stage is the formation of a thin silica blanket on the surface of the droplet. Silica formation begins at the droplet interface where the base meets TEOS. The third stage is the composite particle formation of interpenetrating domains of PE and silica. Any phase separation that does occur, as evident from the existence of crystallinity in the composite, is limited to domains of the order of 20 nm (determined from small-angle X-ray scattering), the size of which is primarily dictated by the polycondensed TEOS. Formation of the initial silica at the interface creates a diffuse zone with both hydrophilic (SiOH) and hydrophobic (SiOEt) residues that facilitate the mixing of the aqueous base with the partially hydrolysed TEOS. As the silica network continues to form, it becomes more and more difficult for the PE to separate, because it is more interpenetrated with the porous inorganic matrix. At the end of this process, the PE and the PE-b-PEG are distributed among several packaging forms: there are domains of crystalline phases (determined from differential scanning calorimetry), separated PE and PE-b-PEG amorphous domains and domains in which the PE and PE-b-PEG are blended. Finally, the organic polymer does not interfere with the silica structure due to the fact that the TEOS polycondensation is a three-dimensional highly cross-linking process, while the linear PE has the flexibility to adapt to the silicate structure.

Slice 1:

Slice 2:

Slice 3:

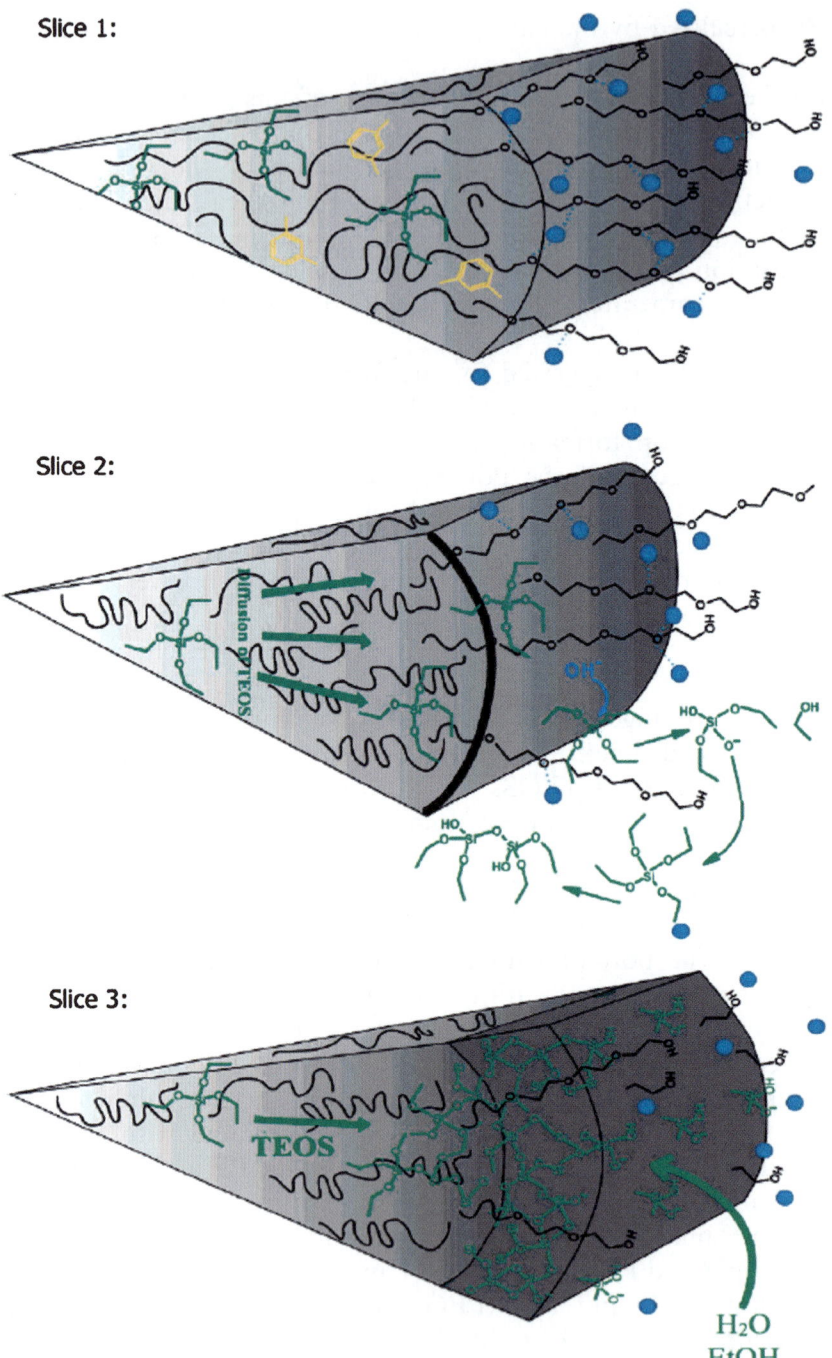

Scheme 7.3 The three stages of the mechanism of formation of the PE@silica particles. (Reproduced from ref. 8, with permission.)

References

1. (a) E. Yilmaz, O. Ramstrom, P. Moller, D. Sanchez and K. Mosbach, *J. Mater. Chem.*, 2002, **12**, 1577; (b) M. Choi, F. Kleitz, D. Liu, H. Y. Lee, W. S. Ahn and R. Ryoo, *J. Am. Chem. Soc.*, 2005, **127**, 1924; (c) W. Y. Chen, K. S. Ho, T.-H. Hsich, F.-C. Chang and Y.-Z. Wang, *Macromol. Rapid Commun.*, 2006, **27**, 452.
2. C. Sanchez, B. Julian, P. Belleville and M. Popall, *J. Mater. Chem.*, 2005, **15**, 3559.
3. G. Philipp and H. Schmidt, *J. Non-Cryst. Solids*, 1984, **63**, 283.
4. A. B. Wojcik, M. J. Matthewson, K. T. Castelino, J. Wojcik and A. Walewskic, *Proc. SPIE*, 2006, **6193**, 61930T.
5. J. Liu and J. C. Berg, *J. Mater. Chem.*, 2007, **17**, 4430.
6. A. Fidalgo, J. P. S. Farinha and J. M. G. Martinho, *Chem. Mater.*, 2007, **19**, 2603.
7. L. M. Ilharco, A. Fidalgo, J. P. S. Farinha, J. M. Gaspar Martinho and M. E. Rosa, *J. Mater. Chem.*, 2007, **17**, 2195.
8. H. Sertchook, H. Elimelech, C. Makarov, R. Khalfin, Y. Cohen, M. Shuster, F. Babonneau and D. Avnir, *J. Am. Chem. Soc.*, 2007, **129**, 98.
9. V. Monteil, J. Stumbaum, R. Thomann and S. Mecking, *Macro-molecules*, 2006, **39**, 2056.
10. (a) K. Tajima, G. Ogawa and T. Aida, *J. Polym. Sci. A: Polym. Chem.*, 2000, **38**, 4821; (b) K. Kageyama, J. Tamazawa and T. Aida, *Science*, 1999, **285**, 2113.

CHAPTER 8

Strategic Aspects of Functional Silicas

8.1 On the Value of Sol–Gel Hybrid Materials

Sol–gel hybrid silicas have plenty of value to offer to practitioners, and society in general. Yet, compared to their potential, their practical applications have only just started to have the impact that in the next ten years will change many things in the chemical and pharmaceutical industry, and thus in practically all manufacturing sectors where chemicals are employed on a large scale. Young chemists applying for jobs in a "hot" area such as sol–gel nanotechnology, and technologists trying to sell R & D to their management, both need strong communication skills to communicate effectively the *value* of these materials.[1]

In general, the growing need for R & D staff is being been fuelled by companies moving towards higher *value* chemical products. Value, however, goes much beyond the simple economic meaning of price. As Lanza del Vasto (Figure 8.1) put it, value is "what is in the man's heart, desire and judgment". In other words, the value of a good is what is in it that is useful to someone. In this precise sense, the pioneer of so-called "lean" manufacturing, Taiichi Ohno (Figure 8.1), defined value as what is useful to the customer and thus something for which customers are willing to *pay*.

Molecular sol–gel encapsulation adds value because, once entrapped in a sol–gel matrix, molecules retain their basic physical and chemical properties that define the practical value of those molecules, but encapsulation also *expands* the properties conferring a whole new set

Silica-Based Materials for Advanced Chemical Applications
By Mario Pagliaro
© Mario Pagliaro 2009
Published by the Royal Society of Chemistry, www.rsc.org

Figure 8.1 Lanza Del Vasto (left) and Taiichi Ohno (right): "value is what is desirable to customers, *i.e.* to man's judgment".

that are "desirable in man's judgment". Pricey oxidation catalyst tetra-*n*-propylammonium perruthenate (TPAP), for instance, dissolved in an organic solvent in the presence of an alcohol, for instance, immediately oxidizes hydrogen peroxide added to the solution. Yet, once entrapped in a SiO_2 matrix, the same catalyst selectively converts the alcohol into carbonyl provided that H_2O_2 is added slowly (Figure 8.2).[2]

From the viewpoint of economic theory, the factors of production—*i.e.* raw materials, labour, knowledge and capital goods—provide *services* which increase the value of a product X relative to the cost per unit of intermediate goods used up in the production of X.

It is enough to visit the clean, small production plant of Sol-Gel Technologies in Israel (Figure 8.3) to recognize that most of the value added to benzoyl peroxide entrapped in microcapsules comes from knowledge—and thus from human ingenuity—which originates the production of the microcapsules. The price at which the white water-based capsule formulation is sold to customers exceeds more than 1000 times the price of the raw materials used to prepare it. Put another way:

"By applying Sol-Gel's technology, pharmaceutical companies can extend and improve current product lines," commented the president of Sol-Gel Technologies Ltd announcing that it has entered into a development and licensing agreement with a leading US pharmaceutical company for the development and commercialization of a major dermatologic product.[3]

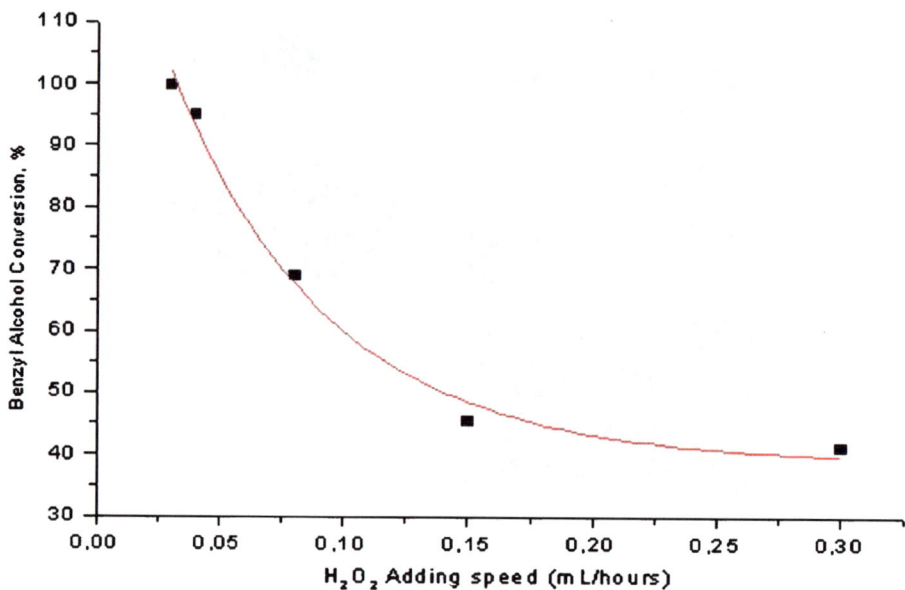

Figure 8.2   Benzyl alcohol conversion as function of the rate of H₂O₂ addition in the presence of silica-entrapped TPAP. (Reproduced from ref. 2, with permission.)

Sol-Gel Technologies

Figure 8.3   Sol–gel entrapment process adds large amounts of value to the entrapped chemicals. And Sol-Gel Technologies was the first company to capitalize on it.

In other words, sol–gel entrapment technology provides innovative drug delivery solutions using sol–gel based encapsulation systems in silica which enables "new and stable combinations of APIs [active pharmaceutical ingredients] resulting in *improved efficacy and usability*".

Again, this improved efficacy resulting from heterogenization of pricey molecules on solid supports has been sought for many years, with far fewer good results, by the fine chemical industry. Indeed, even older and less powerful heterogeneization technologies add significant financial value. For instance, compare the 2007 prices of supported perruthenate (a selective oxidation catalyst) from a common chemicals catalogue. Whereas 40 mmol of perruthenate (as potassium salt) is commercialized at €930,[4] once supported over 100 g of solid, the price becomes €1415 on silica, €2420 on organic resin and €2900 on FibreCat (Figure 8.4).

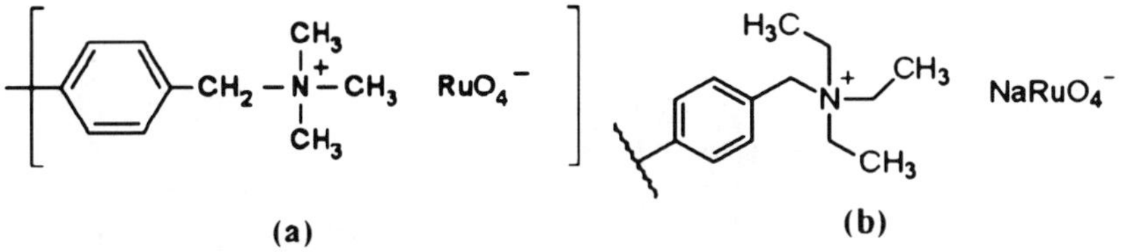

Figure 8.4 Even if supported over solids of poor stability and limited applicability, such as (a) an organic resin, or better (b) inorganic–organic polymer FibreCat, the value of heterogenized perruthenate commercial catalysts is enhanced.

Similar arguments hold for composite silica–polymer glasses of Hybrid Glass Technologies; for the silane-based coatings commercialized by Evonik; and for the nanobinder for paints developed by BASF. In each case, the value of the sol–gel product greatly exceeds that of the components, affording again a whole new *set* of properties, with the products behaving as *multifunctional* materials capable to meet numerous different requirements in a single material. This large value accretion process is what eventually determines their utilization in disparate fields, in optics, electronics, membranes, coatings, catalysis, sensors, biotechnology and medicine.[5]

8.2 Market Trends

In a 2006 report,[6] the global market for sol–gel products, estimated to be $1 billion in 2006, was forecast to increase to $1.4 billion by 2011 with an annual growth rate of 6.3% from 2006 to 2011 (Figure 8.5). The US market for sol–gel ceramics, which at that time in terms of market share was 32% of the world market, was expected to reach $500 million by 2011 with an annual average growth rate of 8.7% per year from 2006 to 2011. Optical and electronic applications are expected to be the fastest growing market segments from 2006 to 2011, at an annual average growth rate ranging from 12 to 15%.

Yet, we argue, the true reason that functionalized sol–gel silicas—developed at an advanced stage already in the 1980s—had to await the first decade of the twenty-first century to reach the market lies in the limited innovation shown by large chemical companies in the 1990s. Prolonged low prices for oil for the whole of the 1990s (averaging $32 per barrel) and the all too common practice of price cartels[7] simply rendered innovation economically not convenient. Unique among other industries, while the chemical industry as a whole is fragmented, the concentration in many product segments is very high with a handful of

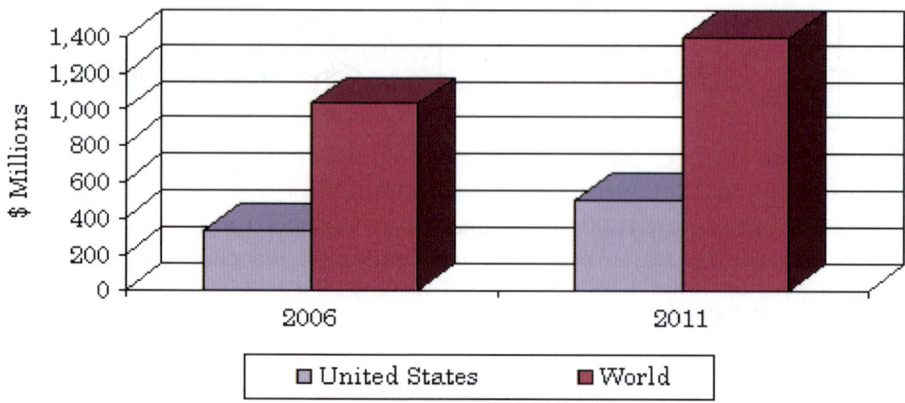

Figure 8.5 US and world markets for sol–gel products, 2006 and 2011. (Reproduced from ref. 1, with permission.)

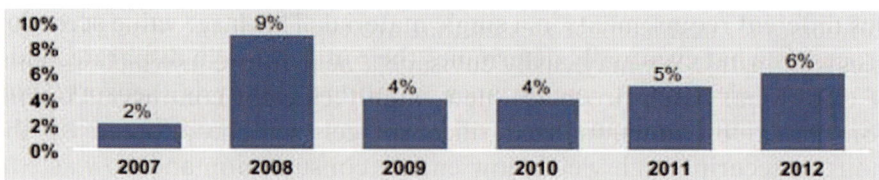

Figure 8.6 Forecast for chemical manufacturing growth in the USA. (Reproduced from First Research, with permission.)

manufacturers holding 80% of that segment.[8] After all, the industry is known for its volatility (Figure 8.6) as demand is driven by the overall performance of many industry sectors, and anything that adds an element of certainty is (and will be) considered welcome to the industry's management.

Subsequent rocketing oil and natural gas prices along with the concomitant action of venture capitalists changed all that. The price of energy and of traditional raw materials suddenly multiplied by a factor three or more. The route to innovative companies in the chemical industry is now open again.

Looking to invest in rapidly growing sectors, venture capitalists have wisely funded the first sol–gel start-ups that are now conquering the markets; remarkably, some of these venture funds are publicly (government) owned as in the case of Canada's SiliCycle. No matter whether private or public, these investments caused the rapid launch on the market of sol–gel functional silicas.

Large, older companies reacted quickly. As late as the middle of 2008, one could read on Degussa (now Evonik) web pages dedicated to its Dynasylan coatings launched in 2007 that:

> Although sol-gel chemistry is known for decades now, examples of successful market introductions are limited. Furthermore almost all existing sol-gel treatments are solvent-based and consequently difficult to handle. As the global leader in water-based sol-gel chemistry, Evonik offers today a broad variety of commercialized water-based products, which are also available in tons-scale.[9]

Yet, the first water-based sol–gel formulations were reported in the 1980s. Perhaps not surprisingly, another large company (BASF) then launched its facade protective nanocoating formulation. Profitability is linked to both manufacturing process efficiency and ownership of innovative products. Sol–gel functionlized silicas serve both requirements. They are generally new, and can be used to streamline all those cumbersome, multistep processes that are typical of the speciality chemicals industry. We may easily envisage a near future in which in place of traditional batch processes, new, highly efficient continuous operations based on functionalized silica sol–gels will be adopted by the industry securing high yields, low energy consumption and no waste to be disposed of.

8.3 Insight into Sol–Gel Nanotechnology

Nanotechnology, namely creating and manipulating objects whose functions are due to dimensions or components at a billionth of a metre scale, has a number of consequences that are truly relevant for tomorrow's global society. And sol–gel functionalized silicas are an eminent class of nanotechnology-based products. In particular, sol–gels are the results of the practice of materials chemistry at the nanoscale, or *nanochemistry*. Like chemistry, this has to do with how we make useful things, as we have learned how to synthesize and exploit new types of materials from individual or groups of nanoscale building-blocks that have been intentionally designed to exhibit useful properties with purposeful function and utility (Figure 8.7).

Indeed, the boundaries that separated these traditional chemistry disciplines in the twentieth century (and chemistry, too, from other disciplines such as physics, biology and engineering) have been broken to create one large multidisciplinary community with a keen scientific

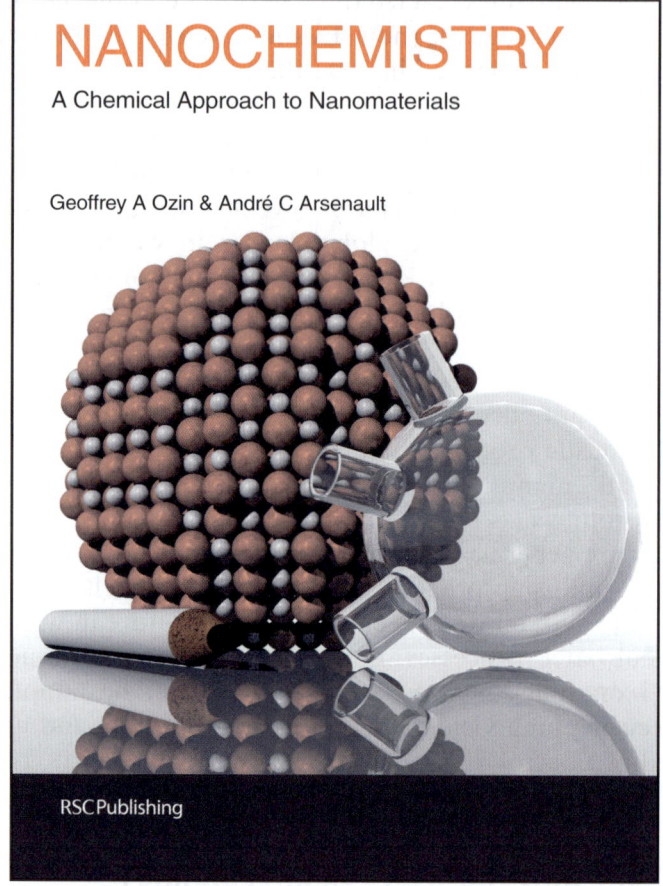

Figure 8.7 Nanochemistry (this is the cover of a leading textbook) is at the frontier of the scientific chemical endeavour of the early twenty-first century.

and technological interest in "all" aspects of the chemistry of materials at the nanoscale. The low temperature and in general the mild conditions of the sol–gel process have caused organic, inorganic and bio-chemistry to merge into a single domain of research (Figure 8.8). In the words of leading materials scientist G. Ozin, "Nano is the beginning of the nanomaterials food chain that will enable and/or disrupt existing technologies or create ones that never existed before".[10]

In practice, as the properties of a (nano)material emerge from the composition, size, shape and surface properties of these individual building-blocks as well as self-assembled architectures made from these building-blocks, chemists are increasingly able to synthesize tailor-made materials from the bottom up. Such techniques generally rely on

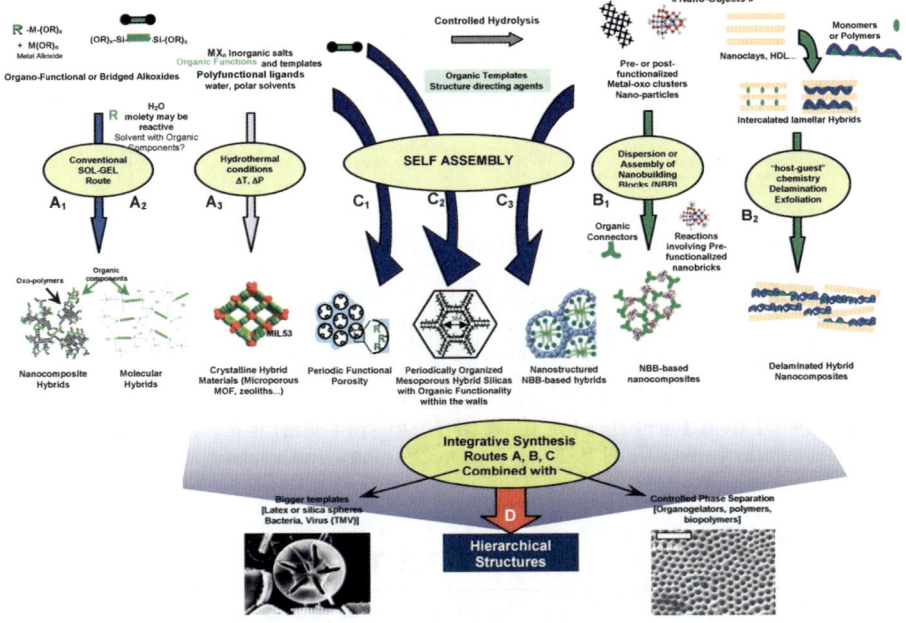

Figure 8.8 Scheme of the main chemical routes for the synthesis of organic–inorganic hybrids. (Reproduced from ref. 1, with permission.)

formulas that control the precise, bottom-up chemical assembly of molecules into geometric structures composed of many molecules. Molecular self-assembly techniques for instance now give us the unprecedented capability of designing and creating nanostructured materials with novel properties.

In his recent *Nano-Hype* Berube[11] concluded that:

> Much of what is sold as "nanotechnology" is in fact a recasting of previous materials science, which is leading to a nanotech industry built solely on selling nanotubes, nanowires, and the like which will "end up with a few suppliers selling low margin products in huge volumes".

Yet, although there has been much hype about the potential applications of nanotechnology, nanomaterials are having a large impact in a number of fields of similarly important societal relevance (Figure 8.9).

Beyond research, powerful trends are already evident in business (Figure 8.10). Not only do a growing number of start-up companies now commercialize products obtained *via* a nanochemistry approach but also national laboratories, military establishments and very big chemical

Figure 8.9 David Berube's blog (nanohype.blogspot.com) is a plentiful resource on the interactions between nanotechnology and society.

The New York Times

Small Business

| WORLD | U.S. | N.Y. / REGION | BUSINESS | TECHNOLOGY | SCIENCE | HEALTH | SPORTS | OPINION |

MEDIA & ADVERTISING   WORLD BUSINESS   **SMALL BUSINESS**   YOUR MONEY   DEALBOOK   MARKETS   RESEAR

ENTREPRENEURIAL EDGE

Nanotechnology Companies Planning to Sell Shares

By JAMES FLANIGAN
Published: December 20, 2007

SIGN IN TO E-MAIL
OR SAVE THIS

NANOTECHNOLOGY companies, nurtured on billions of dollars in government grants and venture investments through most of this decade, are getting ready to go public.

🖨 PRINT

REPRINTS

SHARE

🔍 Enlarge This Image

Being near taking such a step is another stage in the evolution of nanotechnology, the science of materials measured at billionths of a meter or one-500th of a human hair.

ARTICLE TOOLS
SPONSORED BY

Peter DaSilva for The New York Times

Liangbing Hu, a scientist at Unidym, checks a carbon nanotube. Unidym uses nanotechnology to help produce touch screens for cellphones and ATMs.

Experts note that nanotechnology-enabled products are already used in industry.

"There are 200 commercial products in cosmetics, apparel and sporting goods in which nanotechnology plays a role," said Lynn E. Foster, emerging technologies director for the

Figure 8.10 The *New York Times* of 20 December 2007 reported on many nanotechnology companies planning quotation at the Stock Exchange.

companies have entered the field and joined the race for new and exciting nanomaterials. Trends in public and private funding of nano-technology evolve accordingly. Global research funding in the USA, Canada, Japan, China, India, Korea and the EU is in the range of billions of dollars. The future opened up by nanotechnology is coming "sooner than we think".[12]

8.4 Sustainability Aspects of Sol–Gel Materials

If the current trend in commercial nanotechnology ventures continues, we will soon find ourselves with a relatively large nanomateriasls industry. And, as put by Murphy, production of significant quantities of anthropogenically derived nanomaterials will inevitably result in the introduction of these materials to the atmosphere, hydrosphere and biosphere.[13]

Yet, post-industrial societies should not make the same mistakes as were made with unregulated use of chemicals, and nanomaterials that enable nanotechnologies should evolve as instruments of sustainability.[14] It is not a coincidence that "green chemistry" (Figure 8.11) principles have been developed since the mid-1990s, concomitant with advances in nanomaterial synthesis.[15] In fact, as the economic, social and environmental problems associated with the sustainability crisis start to become evident on a global scale, we urgently need advanced nanotechnologies capable of allowing us to drastically reduce emissions and, at the same time, to increase productivity.

Nanomaterials may differ from other particulate materials in both size and surface chemistry; and it is unclear how anthropogenically created materials might differ in their interactions. In particular, the impact of nanoparticle chemistry, and interactions with surfaces are of relevance here. Covering up potentially toxic surface groups with an innocuous organically modified silicate coating improves the biocompatibility of nanomaterials and eliminates safety and environmental risks. For example, cadmium-based quantum dots and gold and silver nanoparticles may all be treated with mercaptopropyltrimethoxysilane ($HS–(CH_2)_3 Si(OCH_3)_3$) resulting in production of an overcoating biocompatible silica shell. Overall, functional organosilica sol–gel materials will act both as

Green Chemistry

Figure 8.11   The principles of "green chemistry" have been spreading simultaneously with advances in nanomaterials synthesis.

safe nanomaterials for a variety of applications, and as those "instruments of sustainability" invoked by Wiesner *et al.*[14]

References

1. R. Arnette, Careers in chemistry, *Science*, 2006, **312**, 1068.
2. S. Campestrini, M. Carraro, U. Tonellàto, M. Pagliaro and R. Ciriminna, Alcohols oxidation with hydrogen peroxide promoted by TPAP-doped ormosils, *Tetrahedron Lett.*, 2004, **45**, 7283.
3. Quoted from the company's website: www.sol-gel.com (as of 20 May 2008).
4. R. Ciriminna, S. Campestrini, M. Carraro and M. Pagliaro, Sol-gel entrapped TPAP: an off-the-shelf catalyst series for the clean oxidation of alcohols, *Curr. Org. Chem.*, 2008, **12**, 257.
5. C. Sanchez, B. Julian, P. Belleville and M. Popall, Applications of hybrid organic–inorganic nanocomposites, *J. Mater. Chem.*, 2005, **15**, 3559.
6. BCC Research, Sol-Gel Processing of Ceramics and Glass, June 2006.
7. For example, P. Meller, Chemical cartel busted, *International Herald Tribune*, 20 January 2005.
8. Cited from: www.hoovers.com.
9. See: www. dynasilan.com.
10. G. Ozin and A. Arsenault, *Nanochemistry*, RSC Publishing, Cambridge, 2005.
11. D. M. Berube, *Nano-Hype: The Truth Behind the Nanotechnology Buzz*, Prometheus Books, New York, 2005.
12. J. Saxton, *Nanotechnology: The Future is Coming Sooner Than You Think*, Joint Economic Committee, US Congress, Washington, DC, 2007.
13. C. J. Murphy, Sustainability as an emerging design criterion in nanoparticle synthesis and applications, *J. Mater. Chem.*, 2008, **18**, 2173.
14. M. R. Wiesner, H. Lecoanet and M. Cortalezzi, Nanomaterials, sustainability, and risk minimization, IWA International Conference on Nano and Microparticles in Water and Wastewater Treatment, Zurich, Switzerland, 22–24 September 2003.
15. J. A. Dahl, B. L. S. Maddux and J. E. Hutchison, Toward greener nanosynthesis, *Chem. Rev.*, 2007, **107**, 2228.

Subject Index

Page references to *figures, tables and text boxes* are shown in *italics*.